2023

ONE THOUSAND SHADES OF GREEN

ONE THOUSAND SHADES OF GREEN

A Year in Search of Britain's Wild Plants

Mike Dilger

BLOOMSBURY WILDLIFE
LONDON • OXFORD • NEW YORK • NEW DELHI • SYDNEY

To the other two corners of my triangle:
Christina and Zachary

BLOOMSBURY WILDLIFE
Bloomsbury Publishing Plc
50 Bedford Square, London, WC1B 3DP, UK
29 Earlsfort Terrace, Dublin 2, Ireland

BLOOMSBURY, BLOOMSBURY WILDLIFE and the Diana logo are trademarks of Bloomsbury Publishing Plc

First published in the United Kingdom 2023

A catalogue record for this book is available from the British Library.

Library of Congress Cataloguing-in-Publication data has been applied for.

ISBN: HB: 978-1-4729-9362-5; ePub: 978-1-4729-9364-9; ePDF: 978-1-4729-9365-6

2 4 6 8 10 9 7 5 3 1

Typeset in Bembo Std by Deanta Global Publishing Services, Chennai, India
Printed and bound in Great Britain by CPI Group (UK) Ltd, Croydon CR0 4YY

Contents

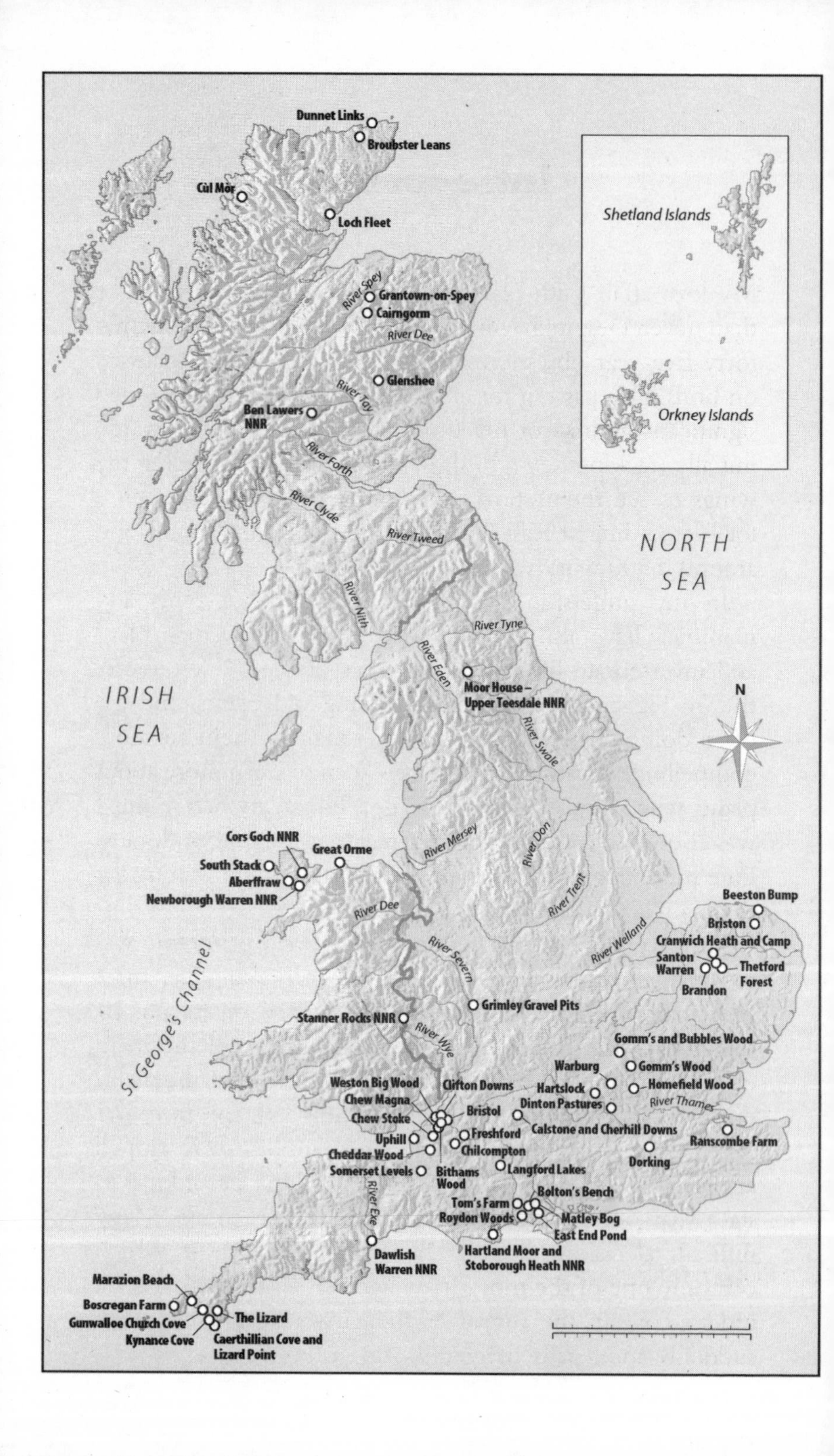
Dunnet Links
Broubster Leans
Cùl Mòr
Loch Fleet
Shetland Islands
Orkney Islands
River Spey
Grantown-on-Spey
Cairngorm
River Dee
Glenshee
River Tay
Ben Lawers NNR
River Forth
River Clyde
River Tweed
NORTH SEA
River Nith
River Tyne
River Eden
Moor House – Upper Teesdale NNR
IRISH SEA
N
River Swale
River Mersey
River Don
Cors Goch NNR
Great Orme
South Stack
Aberffraw
Newborough Warren NNR
River Trent
Beeston Bump
River Dee
Briston
River Welland
Cranwich Heath and Camp
St George's Channel
River Severn
Santon Warren
Thetford Forest
Brandon
Grimley Gravel Pits
Stanner Rocks NNR
River Wye
Gomm's and Bubbles Wood
Warburg
Gomm's Wood
Weston Big Wood
Clifton Downs
Hartslock
Homefield Wood
Chew Magna
Bristol
Dinton Pastures
River Thames
Chew Stoke
Calstone and Cherhill Downs
Freshford
Ranscombe Farm
Uphill
Chilcompton
Cheddar Wood
Dorking
Somerset Levels
Bithams Wood
Langford Lakes
River Exe
Bolton's Bench
Tom's Farm
Roydon Woods
Matley Bog
East End Pond
Dawlish Warren NNR
Hartland Moor and Stoborough Heath NNR
Marazion Beach
Boscregan Farm
The Lizard
Gunwalloe Church Cove
Kynance Cove
Caerthillian Cove and Lizard Point

Introduction

My love affair with plants was a classic case of better late than never. Despite having a university degree in botany, my forty-five-year obsession with wildlife has mostly focussed on birds. This passion for all things feathered has consumed significant chunks of my time from adolescence onwards, initially tracking down birds in Britain before spreading my wings to see them abroad. And along the way I've seen a lot – recording at least 350 different species in Britain, and around 2,500 worldwide.

In my profession as a television naturalist, birds and mammals have also featured more prominently than plant and invertebrate life too. In the eyes of the TV producers they're bigger and sexier than boring old plants, and are often doing something interesting – making them far more compelling to watch. While always keen to learn more about plants when opportunities prevailed, like many before me, I was also guilty of consistently relegating the role of flora to little more than that of a supporting artist, while waiting for the main act to turn up.

However, one consequence of the Covid pandemic was to change this blinkered view of plants in the most spectacular fashion. As the world was tilted on its axis by the spread of a hitherto unknown virus, one of the main effects, certainly in Britain, was the biggest curtailment of our freedom in living memory. As the death toll from the virus began to mount in 2020, and families were ripped apart in the cruellest of fashions, it wouldn't be unkind to state that our government was in full-crisis mode. Many difficult decisions had to be made quickly, of which certainly one of the most draconian was that in a desperate bid to contain the spread of the virus, many of us were suddenly and legally obliged to stay at home.

Being a freelancer, this work-from-home rule had an immediate and disastrous impact on my work. Filming projects, talks and tour-leading trips were suddenly either cancelled or postponed indefinitely, in a fast-moving situation well beyond my control. For many parents however, the real game-changer was the immediate closure of schools. So on top of trying to keep their jobs going from home, they were then required to take on the additional and full-time role of schoolteachers. Being parents ourselves to a seven-year-old boy, this certainly had a seismic effect on my wife Christina and I, as each weekday between 8.45 a.m. and 3.20 p.m. a precarious game ensued, which consisted of juggling what little work we had, while also keeping Zachary up to date with his schoolwork.

I consider myself a 'glass half full' person, and to paraphrase Charles Dickens, it was the worst of times, but it was also the best of times. In a normal year I would expect to spend long periods away from both home and family, by, for example, filming pine martens in the Scottish Highlands one week, before then travelling south to catch up with basking sharks off the coast of Cornwall. But with my regular work grinding to a halt, and all but key workers confined to their homes, the pandemic also presented me with a once-in-a-lifetime opportunity to spend more quality time with my family than ever before. Fortunately 'lockdown' did not in reality translate into 24 hours in solitary confinement, as we were allowed out of our homes to exercise once a day. And for a family who have always made the most of the great outdoors, this proved to be nothing short of a lifeline.

We live in the small Somerset village of Chew Stoke, in the pastoral Chew Valley, which is itself sandwiched between the city of Bristol to the north and Wells to the south. But perhaps the biggest blessing as to where we live is easy access to the wider countryside via a multitude of footpaths. The instruction to stay local and use open spaces close to home proved easy to comply with, as Zachary, border collie

Bramble and I headed out for our daily exercise along any number of different walks, which all started and finished at our back door.

Of course, an obsession with natural history is not something you can simply turn off like a tap, and upon heading out, these daily walks quickly morphed into both green therapy for me and a green classroom for Zachary. Slim, tall for his age and with a tousled mop of sandy-coloured hair, despite us sharing relatively few physical characteristics, personality-wise he was a chip off the old block. Keen to share my enthusiasm with the Young Padawan, I took every opportunity to point out to him the different birds along the way. While expressing polite interest in his father's impromptu ornithology lessons, Zachary appeared more interested in the wild plants, rather than the birds – perhaps because they never flew away.

Many of these first plants that we looked at together were the widespread species I'd known forever, but once the likes of primroses and lesser celandines had been established, the issue of identifying the somewhat trickier speedwells and forget-me-nots, for example, left me unsure and frustrated. Unhappy with this state of affairs, I realised something would have to change. Then just before setting off out for one nature trip I made a momentous decision: Instead of taking out my binoculars as usual, a quick change of mind led me to taking out a wildflower field guide. This change of emphasis was simple but brilliant, and effectively took the handbrake off my botanising. Everybody loves to learn, and by occasionally stopping to identify any plant whose identity I was unsure of – and which I'd normally have strolled right past – gave me that similar euphoric feeling I'd experienced when first learning about birds as a kid.

Whether it be birds or beetles, I've always enjoyed the process of using my knowledge and deductive reasoning to identify wildlife. When systematically working through the key features of what you suspect to be a certain species and suddenly realising your assumption was correct, it is nothing short of thrilling. I would go so far as to say that I can only properly enjoy watching a certain species when I know what it is that I'm actually looking at. This ability to ascribe a name to plants and animals can be really empowering, as this crucial label can then be used as a tool with which to access any amount of easily accessible research about that species.

Also, at a much more basic level, who doesn't enjoy encountering something they've never seen before? I clearly remember seeing my first ever golden eagle, for example, like it was yesterday, and the ensuing buzz that comes from clapping eyes on a 'first' certainly saw me chasing all over the United Kingdom to 'tick' new birds in days gone by. I'm not alone by any means in the pursuit of rarity either, with many birders spending their whole lives – and a sizeable portion of their disposable income – constantly chasing the next rare bird to turn up.

Even though the flame of bird-chasing had long since been snuffed out by the demands of a busy job and a family, this compulsive need to see new species has never gone away, and perhaps just needed a new direction. So while our daily Covid walks offered me little opportunity to see any new birds, it was a very different story for any number of the plants we encountered. As we pounded every possible path in the Chew Valley during that unforgettable spring of 2020, it almost felt that the turn of every corner might present the opportunity of not only solving a taxonomic conundrum but also potentially being rewarded with the ultimate prize: that of a shiny, new species.

Apart from the obvious pleasure of finding new plants on our walks, there was a bigger force at play, whereby the process of learning more about botany was also making both father and son better naturalists. I meet many birders in both my personal and professional life who, despite being astonishingly talented ornithologists, are simply unable to identify even the most ubiquitous of plants. But having reignited my own zest for botany, I was now able to see what folly this was. Plants are clearly the life-support systems for all terrestrial life, and without them at the base of every food pyramid, everything would quickly crumble.

As someone with a Calvinist work ethic ingrained, I'm generally happiest when busy, and with more time on my hands than I'd ever experienced in my adulthood I also needed a project to get my teeth into. In just a short time, and by placing plants front and centre, I quickly developed the zeal of a convert. Buying into the phrase 'to know them is to love them', I was also determined to use my profile to raise awareness, and wanted my next big mission to be that of 'Plant Champion'. I had known for a long time that wild plants (but not cultivated garden plants) were neglected by the mainstream media, to the extent that in over 400 short films I've made for BBC's *The One Show* as their wildlife reporter, no more than half a dozen had focussed solely on plants.

For some reason, all my best ideas tend to come to me in the shower. It's a time perhaps when my mind is at its least cluttered and I'm guaranteed to be left alone. On one such morning in the autumn of 2020, while preparing for another day's home-schooling, the idea of a big botanical year suddenly came to me. Why couldn't I spend the whole of 2021 trying to see as many of Britain's plants as possible? Being a 'lister' it seemed like I would need a target to aim for, but attempting to see all 2,412 plants currently listed in 2002's *New Atlas of the British and Irish Flora* was as ridiculous as it was impossible. However, with a whole lot of planning, hard work and single-minded dedication then perhaps a figure of 1,000 might just be achievable? Also, being confronted with all manner of new species would test my identification skills like never before. So when singling out sedges, for example, I would need to get up to speed – and quickly.

Once I'm enthused by an idea, I can often – to my family's detriment – be like a dog with a bone. Keen to keep my big botanical year as carbon-efficient as possible, which was only sensible given the impending climate emergency, finding many of the plants close to home would be imperative. This meant botanically important locations on my doorstep, such as the Mendips and the Avon Gorge, would need to feature heavily. But providing we would eventually be allowed to travel further afield, I figured there might also be the opportunity to fit in a few visits to exciting botanical locations, particularly if I could piggyback them onto other commitments. Furthermore, it would be desirable to see the plants when looking at their best. Birdwatchers, for example, always prefer to see their birds when in breeding plumage, and so by transferring this sentiment to a botanical context, I would be keen to see as many of the plants as possible when looking at their finest – in other words, when in flower.

Having spent 15 years on *The One Show*, I've been in the privileged position to probably travel to more remote locations around Britain than even Sir David Attenborough. But there were still some places that I also needed an excuse to visit, such as the fabled botanical locations of Teesdale and Ben Lawers for example. The fact that settings like these hold some of our finest botanical jewels, but comparatively little else, is perhaps the main reason why camera crews so rarely go there. It was not just these locations I was keen to visit; there were certain plants I also wanted to see too. Spring gentian, for example, has to be one of our most stunning plants, and, being confined to Teesdale's upland grasslands, must also be one of our rarest.

Rare plants undoubtedly have a cachet, and although while compiling my list all species would be treated as equal – with bluebell and lady's slipper orchid only counting as one point each, for example, some plants were definitely more equal than others. I would be lying if I didn't declare that clapping eyes on some of our rarest plants was a large part of the project's attraction, but I could also use this golden opportunity to learn more about the lives of these plants. Many rare plants are rare for a reason, and my mission, were I to accept it, would also allow me to personally gauge how our most threatened plants were coping during an era of frightening habitat degradation and fragmentation, and at a time of unprecedented climate change.

Having convinced myself that I wanted to undertake a big botanical year, I now had to convince my wife and son. With a family comes responsibility, and if I were to have any chance of succeeding I would need two things: their blessing and, where possible, their involvement. Calling a family summit, I pitched my hare-brained idea and explained how their respective contributions would be invaluable. I argued that they both adored being outside, and with Zachary loving plants and Christina working with them in her capacity as a gardener and botanical illustrator, what was not to like?

And I think you can guess what they said …

CHAPTER 1

New Year Plant Hunt

The Christmas tree was still up, but I couldn't contain myself; I was champing at the bit.

Having decided to make 2021 my 'big botanical year', I needed to start January with a statement – a big botanical bang. But with the winter solstice still fresh in the memory, one might expect floral pickings to be a touch on the slim side at this time of year. Thanks, however, to the seemingly unstoppable juggernaut of climate change, hunting for flowering plants early in the new year is now nowhere near as daft as it once perhaps sounded.

The United Kingdom's meteorologists are of an accord that our most recent couple of decades have been warmer, wetter and sunnier than those even in the latter part of the twentieth century, with this pattern highly likely to continue along the same trajectory, unless we change our ways. Despite the potential impacts of a changing climate on our flora still poorly understood, anyone spending any time out of doors will undoubtedly have realised that as white Christmases are becoming increasingly relegated to the dim and distant past, green New Years are now seemingly the norm.

Plants are already responding to these changing climatic conditions, with a recent report suggesting that many British plants are now flowering a month earlier, while autumn leaf fall is also being delayed. One wildlife charity keen to both quantify and highlight these changes is the Botanical Society of Britain and Ireland (BSBI). Flowering plants can certainly be considered canaries in the coal mine when it comes to monitoring changes in the natural world, as their inability to simply up-sticks in the face of less favourable conditions puts them firmly in the 'lump it' category. It was with this in mind that the BSBI launched their inaugural New Year Plant Hunt in 2002. This citizen science initiative involves persuading as many of its members as possible across the United Kingdom to take a three-hour walk early in the New Year, while noting any flowering species along the way. If I too were to join this enterprise it would also provide the perfect opportunity to hit the ground running. As an added incentive, any data I collected would also contribute to the much larger collective effort aimed at better understanding how our plants are coping in a changing world.

Certainly with my university degree in botany, learning to identify wild plants had bizarrely never been part of the syllabus, which meant that my knowledge of British plants was limited to little more than a familiarity with our commonest fifty or so species. And so the relatively small suite of plants which were capable of flowering at this time of year would also mean I wouldn't in turn become dispirited by the number of species I was simply unable to identify. Another advantage of taking part in the hunt was the availability of expert online help, so any flowering plant I didn't know out in the field could hopefully be photographed for identification back at home.

Accompanying me on this very first excursion would be my trusty copy of *Wild Flowers of Britain & Ireland*, by Blamey, Fitter and Fitter (BFF), a botanical eye lens and my immediate family, comprising Christina, Zachary and Bramble. Early on in the extensive planning process I had realised that while some of the outings would inevitably have to be solo adventures, I was keen to turn as many as possible into a family affair. Of course, having my nearest and dearest perennially in tow while my head was constantly in the herbage could end up being a double-edged sword, but having weighed up the pros and cons of their involvement, I eventually came to the conclusion they might just be one of my best assets.

My wife Christina, for example, also has a degree in biological sciences. Ordinarily this might be considered advantageous, but as field botany is currently held in such low regard, many students can graduate with a first-class honours degree in biological sciences without even being able to distinguish a dandelion from a daisy. We met each other on the BBC's first ever *Springwatch*, and although not as obsessed a naturalist as me, Christina's selling point undoubtedly came from her subsequent (and current) job as a Royal Horticultural Society (RHS)-qualified freelance gardener. As such, my green-fingered wife had slowly but surely acquired a vast knowledge of horticultural plants, putting her in the frame to provide some much-needed consultancy when identifying, for example, any cultivated plants that had escaped the confines of the garden.

Zachary, on the other hand, brought a different set of qualities to the table. Having spent much of 2020's various lockdowns as my constant companion on our daily dog walks, his presence had been instrumental in rekindling my love of botany. It was only with Zachary, for example, that I'd finally bothered to get to grips with the precise identities of the violets and forget-me-nots that I had been blithely and ignorantly walking past for years. Also, like I was at his age, he thought being indoors was dull, when compared to all the fun to be had out of doors. Being so young also meant he had that capacity for soaking up information like a sponge, and – perhaps most importantly of all – genuinely seemed interested in learning more about plants.

Furthermore, one unfortunate manifestation of my advancing years was my increasing reliance on my varifocal glasses, not only for finding the plants in the first place, but also for helping to observe the salient – and often small

– features necessary for differentiating between closely related species. So if I could harness Zachary's visual acuity, while channelling his boundless enthusiasm, then he might just prove a valuable trump card.

Our final member of the party would be our three-year-old border collie Bramble, whose main contribution would be to encourage us out in the first place. The only other help our canine companion might conceivably offer would be through his remarkable propensity for finding either the most interesting or rarest plant at any one site upon which to urinate.

The plan had been to assemble our motley crew for a circular stroll from home, taking in both a slice of countryside and the delightful village of Chew Magna along the way. Ideally 1st January would have been the most obvious choice for the commencement of the mission, with the added benefit of a long walk also burning off any excess calories laid down over the festive period. But with the weather such a damp squib during the first few days of the year, a delay to 4th January, when sunnier weather was predicted, would not only make the hunt a more pleasant experience, but might also encourage a few more early-flowering plants to tentatively unfurl their petals.

Living at the bottom of a small cul-de-sac, we must have been no more than a couple of minutes into our hunt, and no further than 30m from our front door, before we were able to add our very first flower to the list. Common groundsel is a weed so spectacularly undistinguished that it can be perfectly summed up by its scientific name of *Senecio vulgaris*. Seeing it sprouting up from the merest crack between the pavement and our neighbours' low garden wall, I was able to instantly appreciate why groundsel derives from an Old English meaning of 'ground-swallower'. Possessing long, lobed and ragged leaves that became progressively stalked further down the stem, the groundsel was topped by an open cluster of flowers devoid of rayed petals (such as the white petals on a daisy for example), with the minimal, yet welcome, splash of yellow provided by the flowers' central disk petals.

Dropping to ground level is always the best way to appreciate any plant like this, and it was not until placing my nose just a smidge away from the plant's modest blooms that I then noticed perhaps its most arresting feature – black tips to the thin green bracts clustered around the base of each flower. This neat little feature made these green leafy scales look like they'd just been freshly dipped in the world's smallest ink pot. My trusty Blamey, Fitter and Fitter (BFF) stated that groundsel was one of a select group of plants to have been 'observed flowering in every month of the year'. In fact so ubiquitous is this annual weed that it had taken the position of second on the BSBI's list of most commonly recorded species on the previous year's plant hunt. So it was in essence an inauspicious plant with which to start my auspicious list. Snapping a couple of pictures of the plant with my phone, I somewhat superfluously announced 'One down, nine hundred and ninety-nine to find,' to my assembled crack team, but Christina's response was that grubbing around a neighbour's wall could be misconstrued as voyeurism, so it might be best not to savour the moment for too long.

As we exited our street, the grassy verge sandwiched in between a wonderfully mature beech hedge and the main road was to provide the home for our second species – a real cracker. Lesser celandine is one of the classic heralds of spring, and quite a sight to behold when hundreds of glossy yellow blooms seemingly float above a green sea of kidney-shaped leaves, while tracking the sun's movements across the sky. These mini satellite dishes are most commonly seen in their massed ranks from late February, with the famous naturalist Gilbert White noting the average first flowering of lesser celandine around his Hampshire village of Selbourne being 21st February. However, in sunny and south-facing locations, a few outliers will often raise their heads above the parapet early, and to my delight the one and only bloom we encountered was found by Zachary. Perhaps even more significantly, despite the fact that he'd obviously not seen it in flower for the best part of eight months, which is statistically a large chunk of a seven-year-old's life, he was still able to recall its name.

Despite being a near-constant presence in grasslands and along hedgerows and woodland edges across most of the United Kingdom, familiarity has not led to contempt, with lesser celandine surely ranking as one of our most celebrated flowers. William Wordsworth, for example, is perhaps best known for his poem, 'I Wandered Lonely as a Cloud', leading to many people incorrectly presuming the daffodil to be his favourite flower. However, the poet held the lesser celandine in higher esteem. His adoration of the lesser celandine led to the decision, upon his death in 1850, to depict the plant on his tombstone. However, in an unfortunate twist of fate, the plant that was accidentally and ineptly chiselled on his monument was the greater celandine, which isn't even a buttercup, but a member of the poppy family!

With the hedgerow keeping the worst of the wind at bay, and a southerly aspect helping to raise the temperature beyond that of the immediate surroundings, it was of no surprise to then find yet another flowering plant thriving a little further along the same stretch. Dog's mercury is probably one of the commonest plants you've never heard of. Possessing spear-shaped, toothed and fresh-green leaves, from the bases of which emanate small and nondescript sprays of yellowy-green flowers, this seemingly harmless plant is, however, capable of thuggish tendencies. Spreading by underground rhizomes to form large clonal patches, this plant has developed the ability to grow in such dense, leafy carpets that it can often smother and ultimately shade out more light-demanding and undoubtedly sexier springtime species, like oxlip and fly orchid.

Furthermore, the plant has also been able to use and abuse old and mature hedgerows, such as the one we were currently looking at, in order to break the shackles of its traditional wooded home. By simply branching out along these hedgerow corridors, dog's mercury has been able to slowly colonise a whole host of new sites. This certainly wouldn't be the last time I'd encounter this botanical wolf in sheep's clothing during my big botanical year.

The Two Rivers Trail is a public footpath to the south of the village of Chew Magna. It joins Strode Brook – the babbling stream which also happens to run along the bottom of our garden – with the somewhat larger Chew River, whose origins lie in the limey Mendips. It was a trail we'd got to know well during lockdown, with a large portion of the walk taking us through mature farmland, which mostly consisted of differently-sized and shaped fields that were either parcelled up by fence lines or ancient hedgerows. Gazing forlornly down on these shorn and gappy hedges for the first time in the calendar year, but the umpteenth time since moving to the Chew Valley, I was struck once again by the limited wildlife interest of much of England's farmland. As agricultural technology has advanced, the dying art of hedge-laying – with its miraculous ability to rejuvenate hedgerows – has been replaced by the brutality of flail-cutting. Now the default method for keeping hedgerows in check, this highly-mechanised technique is little more than death by a thousand cuts, as each of these living boundaries is treated to a quick short back and sides.

The farmers are, of course, worried that as the hedgerows grow upwards and outwards, they will diminish the land available for cropping and grazing, which will in turn cut their profit margins. But the end result of repeatedly nipping back hedges to the same point is to ultimately reduce them to little more than gnarled, unhealthy stumps, which are of limited interest to any wildlife searching for food or shelter. And as the hedgerows become increasingly ineffective as livestock barriers, this will ultimately lead to extra expense for the farmer as a fencing contractor has to be employed.

Away from their brow-beaten perimeters, the fields themselves consisted of a mix of improved grassland pasture and arable fields, and while subtly different habitats to the trained eye, they shared the depressing trait of offering few opportunities for wildflowers. Agricultural weeds must be amongst the most besieged of all plant groups in Britain's intensively managed countryside, where the days of seeing fields of red, blue and yellow from common poppies, cornflowers and corn marigolds have long since gone. These colourful splashes in the farmed countryside have in essence become overwhelmed by the massive agricultural revolution. As some farmers have adopted more efficient seed-cleaning techniques, used ever higher levels of herbicides, increased nitrogen applications and mechanised the whole process, the end result has been an unfair fight, with weeds the inevitable loser. Ever keen to champion the underdog, I found agricultural weeds mercurial and enigmatic as a group, and as such wanted them to be well represented on my list.

Standing in one such field just south of Chew Magna, which had previously been used for cultivating maize, I reckoned this would be my first and best opportunity to bag a few members of this downtrodden faction of the nation's flora. The good news, however, was that the field was marginally more productive than I'd originally thought, and in quick succession we were able to add common field speedwell, red dead-nettle, scentless mayweed and common radish to a list that was by now steadily galloping towards double figures.

Potentially confusable with a couple of closely related species, only the scentless mayweed needed a quick consultation from my BFF before being satisfied that my tentative and initial identification had indeed been correct. Moreover, the tattered state of the radish and mayweed plants suggested that these were species very much at the tail end of their flowering season – doubtless helped by the consensus that modern winters now appear intent on becoming the 'new autumn'. By contrast the dead-nettle and speedwell flowers looked far fresher, indicating they'd kicked off their floral season at about the same time that we'd been ringing in the New Year.

Leaving the field, and with a sense that I was finally hitting my straps, we wandered towards Chew Magna's high street, which is reached by a couple of meandering residential lanes lined with tall, lanky hedgerows. All members of the newly established 'Chew Stoke Botanical Group' were by now making fulsome contributions, as unseasonably early flowering representatives of hogweed, primrose and wood avens were all added to the day's inventory. Despite a nagging feeling that we were gorging on the botanical equivalent of low-hanging fruit, and more challenging trips would surely lie ahead, it nevertheless felt splendid to get off to such a good start.

Upon encountering a small and weedy crucifer growing from the base of a wall, I suddenly realised I'd come across my first identification dilemma. I'd always found members of the cabbage family (or *Cruciferae*) difficult to get to grips with, as the group possesses a relatively large number of similar-looking species. Deciding the plant is a crucifer in the first place is perhaps the easiest part, as the flowers of virtually every signed-up member of this family possess four petals arranged in a distinct cross or cruciform pattern. Further subdivision, however, then quickly becomes far trickier, especially if the identifier happens to be a botanical novice like me.

Calling over Christina, I declared my suspicions that it might be one of the bittercresses, but beyond this was unsure as to its precise identity. With barely a glance my wife instantly and assuredly replied, 'Easy … that's hairy bittercress.' Seeing the incredulity written across my face, she went on to explain that this weed was so omnipresent in virtually every herbaceous border she'd ever maintained that it was as familiar as the back of her own hand. Having already accepted that I probably wasn't the best botanist in the Chew Valley, I might additionally have to concede that I wasn't even the best botanist in my own household!

As we moved into the heart of the village, Chew Magna's numerous retaining walls and grassy verges, which would hopefully comprise a suite of species more familiar to me than those in the 'garden weed' category, became the next focus of our attention. I had spent a good deal of time poring over the BSBI's online checklists compiled from previous years' hunts, and a fact had arisen from my research that was suddenly and retrospectively blindingly obvious – urban areas floristically outperform all other habitats at this time of year.

The two main reasons urban areas are so botanically rich in the dead of winter are not particularly difficult to fathom when you think about it, with the most obvious being their close proximity to gardens. As a keen gardening nation, Britain's walls, pavements and gutters have become awash with species that have simply hopped over the garden wall. The second reason is down to the 'Urban Heat Island' effect. This phenomenon is caused by the ability of pavements and buildings to absorb and retain heat far more efficiently than the surrounding countryside, with research in London, for example, revealing that intensely urban regions have been recorded as being up to 10°C warmer than their neighbouring rural districts.

Some of the escapees, such as red valerian, ivy-leaved toadflax and the delightful Mexican fleabane, were species

I'd seen for years. These were not only familiar fixtures on many West Country walls, but also seemed to be plants in the groundsel category – or with a January–December flowering season. Once added to the list these species became effectively discarded, as my focus then switched to finding two species of bellflower I'd previously seen in the village. Known by botanists as 'port' and 'posh' due to their virtually unpronounceable species names, Adria and trailing bellflower both originate from the Balkan region, and in recent decades have become colourful colonisers of cracks and crevices in and around Bristol's walls. Providing a welcome purple haze right through spring and summer, these were two species that up until now I'd never taken the time to properly differentiate. Calling me over to a section of wall close to the village's primary school, Zachary was keen to show me two discreet flowering patches he'd located, that even to the untrained eye looked like two distinct species. But which was which?

Certainly one appeared more purplish-blue while the other seemed more bluish-purple, but the devil, as I was rapidly discovering on my crash course in field botany, was in the detail. After a considerable amount of time comparing and contrasting, which then necessitated a quick consultation of my BFF, the differences between these two remarkably similar species suddenly became strikingly obvious. 'Posh' had noticeably longer petals, and when viewed from above looked more like a star, while the more funnel-shaped 'port' had more ovate-shaped petals, which were cut just a quarter back to the base. What teamwork – the plants were found by the son and subsequently identified by his dad.

Perhaps this ridiculously ambitious project might just be achievable after all.

CHAPTER 2

Lockdown Botany

Returning home with a feeling of quiet satisfaction that my challenge was finally underway, little did I know that it would only be a few short hours later that a normally ebullient Boris Johnson would gravely inform us during a state-of-the-nation broadcast that the United Kingdom would have to enter a third lockdown.

Apparently a new Covid variant was spreading so quickly and uncontrollably that the only course of action left for the government was to close schools with immediate effect and confine us to barracks. As the blood drained from my face at this seismic news, two immediate thoughts crossed my mind: Zachary being at home would seriously hamper his parents' ability to work and also my carefully laid plans of wall-to-wall botany had suddenly become far more difficult, if not impossible. On the one positive side, this extreme measure would be undertaken at a time that was still relatively quiet for both my broadcasting work and plant-hunting. We would just have to hope that this short, sharp shock would keep the virus sufficiently in check until finer weather brought more freedom in the spring.

Nevertheless the following month dragged along at a snail's pace as Christina and I were reminded of how difficult home-schooling actually is. Apart from time spent in the garden, the only other periods outside were the short circular walks we were allowed as 'permitted exercise', which invariably started and finished at our back door.

Hearing reports of people being prosecuted for simply having had the audacity to walk in the countryside certainly made us nervous of travelling further afield, especially as the rules and regulations surrounding what constituted 'exercise' were ambiguous at best. Keen to clear up any confusion, 10 Downing Street then advised that Britons would be permitted to visit any location, providing exercise time was double that of the time taken to drive there. Sensing I could use this clause to get my foot in (or out of) the door, I decided that if we were to visit a nature reserve a mere 28 minutes' drive away, all we would need to do to show compliance would be to spend 56 minutes outside taking in lungfuls of fresh, cold and Covid-free air … piece of cake!

I consider myself incredibly fortunate to have a family that never has to be cajoled or persuaded outdoors. Christina's childhood was one of long family walks come the weekend and Zachary had always preferred tree-climbing to screen-timing. In fact even before both Covid and the big botanical year's inception, being outside had quickly become our family's default option whenever we wanted to spend quality time together. So when I suggested a wintry woodland walk that crucially didn't appear to break any of the rules, Christina and Zachary both leapt at the idea.

As any parent and dog-owner will testify, sometimes even just getting out of the house can be a triumph in itself. The much-publicised 'Beast from the East II' was apparently on its way, so with a cold wind guaranteed, warm clothing and reasonable footwear (times three) had first to be packed in the car. Having opted for a picnic at the site, which would ensure we'd sail well over the 56-minute mark, the preparation of sandwiches and filling of flasks then became the next time-consuming job to tackle. Our son had recently become obsessed by the 'derring-do' of a certain Bear Grylls, which meant that he would also need time to pack a penknife, rope, carabiners and stove – because you never know when a gentle stroll around the woods might suddenly turn into a full-on survival weekend. And by the time all the doggy paraphernalia had been added to the 'to go' pile, we probably had enough supplies to last for a month in a nuclear bunker.

Our destination was Weston Big Wood, a 42-hectare wood managed by the Avon Wildlife Trust and considered one of the richest woodlands for wildlife in what is now a defunct county. Purchased by the Trust shortly after the abolishment of Avon in 1996, the reserve itself consists of a narrow ridge, with steep-facing flanks, which effectively connects the towns of Portishead and Clevedon, while also separating the Gordano Valley from the Bristol Channel.

The presence of two imposing and abandoned quarries flanking the reserve's southerly and westerly boundaries represents the only clue needed when guessing why the Trust had felt compelled to step in. The ridge is still actively mined for limestone, which is used as aggregate by the road-building industry, and the fear in 1996 had been that the bedrock on which the woodland was perched might at some unspecified date in the future have been deemed fair game. So the fact that the limestone underlying the reserve's rich loamy soil would now remain where it had originally been laid down (some 360 million) years ago, was great news, not just for the reserve's admiring human visitors, but also for its wild inhabitants – such as the plants.

Of the many reserves within a half-hour's reach of home, Weston Big Wood was not just a fine example of a southern limestone woodland, but also contained a couple of spring botanical specialities that I was keen to add to the list. Despite both possessing 'spurge' in their titles, this really is the only common ground that wood spurge and spurge-laurel share, as the former is actually a member of the spurge family, while the latter is one of just two British members of the far more unusual mezereon family.

The woodland itself is believed to date right back to pre-Roman times, as evidenced by a range of ancient woodland indicator plants, such as wild garlic, bluebell and wood anemone. But given the early timing of our visit, there would be little evidence of these classic harbingers of spring, meaning the two target species would hopefully be even easier to spot on the sparse woodland floor. The reserve's online blurb also claimed the site to be 'well-drained', but we found, upon entering through the kissing gate, that the paths were depressingly similar in appearance to the clay-sodden and claggy ones we'd left behind in the Chew Valley. Obviously even the reserve's porous limestone had been unable to cope with both the amount of rain we'd recently received and the increased footfall that all reserves close to conurbations had contended with during the various lockdowns.

One of the best aspects of many of the hundreds of Wildlife Trusts' reserves up and down the country are their excellent interpretive facilities. And after absorbing the contents of a particularly welcoming and informative board, we decided to take up their suggestion of a circular walk of around 90 minutes, that would take in many of the reserve's best locations along the way. The first section involved a muddy and zig-zagged ascent through a glade of what appeared to be small-leaved lime trees. Once widespread in many of England's woods, this native tree has since become far trickier to find due to the proliferation of a domineering hybrid. Created by the small-leaved's promiscuity with our other native lime – the large-leaved species – the ungrateful love-child that is the European lime has been doing its level best ever since to bury its parents under the patio.

Aware that buds and bark are key to clinching any tree identifications so early in the year, I was soon able to locate a suitably accessible and likely-looking candidate close to the path. In small-leaved lime the buds possess two chestnut to wine-red coloured bud scales, with the lowest bud scale marginally more than half the height of the bud itself. And confirming this to be the case here, I then adopted the belt and braces approach to identification by additionally examining the tree's bark. According to the literature I'd absorbed beforehand, small-leaved lime's bark should be 'longitudinally fissured and deeply ribbed'... which it was. Hey presto, I'd officially added the first tree species to my embryonic list.

In an ideal world I would have liked to have recorded most species when looking their best, such as plants in flower and trees in leaf, but being only too aware that May, June and July would be the most frenetic period of identification, getting some of the trickier species out of the way was no bad idea. I'd already decided that adding ferns to the list would be more than acceptable, providing I was of course 100 per cent confident with my identification in what can be a decidedly difficult group. Unlike our deciduous trees and most woodland herbs, hardy ferns can often look their best in winter, and provide a fabulous and much-needed dash of green to counteract the greys, browns and whites predominating at this time of year. A prime example surrounded us in the form of the strap-shaped leaves of hart's tongue, surely vying with bracken for the mantle of Britain's most easily recognised fern. A common presence across many of our woodlands and walls, the fern's glossy green fronds tend to be at their most dominant wherever limes are present, as was patently the case here.

While the ubiquitous hart's tongue fern would surely represent one of the most straightforward of fern additions to my list, the precise identity of the fern plastering the mossy boughs just above our heads was another matter entirely. Living an epiphytic lifestyle – which is when a plant grows on another plant to gain access to higher light and moisture levels – I had previously identified this distinct and elevated fern as being a species called common polypody. However, it was only during my winter preparations that I had subsequently discovered that this fern had recently been split by a panel of botanical pedants into three distinct species. If that wasn't enough, the waters had become muddied even further, with the discovery that two of the new polypody species had pulled off the same trick as the lime trees – by getting together to produce a hybrid. Despite being able to discount one of the three full species and the hybrid due to their undoubted preference for an acidic geology, ascertaining whether it was either southern or intermediate polypody was, I decided, way above and beyond my current abilities. Apparently these two very similar species are split on subtle differences in frond shape and on the colour of the annuli (or walls) of the spore capsules on the underside. But with my family imploring me to put my book away, as they were getting cold while I stared at spores, I had to leave the plant without an identification clinched. I realised that it wouldn't be the last time I'd just have to walk away, as I caught up with my family further along the path.

Concentrating now on the understorey, or plants closer to the woodland floor, as we walked to warm up, I consoled myself with a couple of staples perhaps best known from a Christmas carol. Holly and ivy must surely figure in Britain's top-ten list of most commonly recognised plants, with their omnipresence possibly conferring a certain dullness. But they are, in fact, anything but. Holly, of course, possesses two leaf forms, with the leaves closer to the ground packed with prickles to protect them from browsing deer. However, a couple of metres higher up and crucially above the browse line, the leaves become smoother-edged, or 'slike', as the production of prickles is deemed an unnecessary waste of resources. And if even that isn't interesting enough to convince you of the plant's worth, holly is also dioecious, meaning that each bush will be either male or female. Sexing your hollies in winter is normally a straightforward process, as only the females possess the bright red berries, but by the late stages of winter any marauding flocks of redwing and fieldfare may well have relieved most of the females of their fruit, in turn nullifying this one sexy standout feature.

By contrast, the berries were still very much in evidence on the ivy in the woodland, our only evergreen liana (a name given to long-stemmed, woody vines), as it gamely attempted to coat the trunks of virtually any tree of substance. Always keen to promote its social standing, ivy must surely be the most upwardly mobile of all our woodland plants. Often initially seen creeping across the woodland floor, ivy cannot resist the temptation to race vertically for the light when the opportunity arises. Using trees as a scaffold, which it adheres to with the help of special adventitious rootlets, ivy is assumed by many to be a parasite with death by smothering on its mind. However, this is a fallacy, as the only competition comes in the form of a fair fight for soil nutrients and water down below, while both host and uninvited guest will also compete for the sunlight up in the canopy. The only trees likely to suffer physical damage directly from the ivy's presence will be those mature, senescing (aging) specimens, where weakened or dead branches may struggle to bear the extra weight on particularly windy days. Mature ivy specimens themselves, however, are capable of living to the ripe old age of a hundred, with their lattice-like structures providing homes for innumerable woodland invertebrates. And when ivy's nectar and berry contributions are added into the mix, it is hard to think of a more wildlife-friendly plant.

As we tramped further away from the entrance, the paths steadily became firmer underfoot and a large patch of wood spurge was quickly located where some recent coppicing had brought light down to the woodland floor. The surprising presence of so many lush evergreen plants, however, would make our task of picking out the dark glossy-green leaves of spurge-laurel perhaps more difficult than I had initially appreciated. To make matters worse, Zachary and Bramble's patience with me constantly stopping to stare at plants was beginning to wear thin. Certainly for our son, spotting trees to put a name to had suddenly become far less interesting than identifying the ones he could potentially climb. But with Christina's assistance in crowd control, running both activities concurrently still looked manageable for a while longer before the plant-spotting would have to be firmly placed on the back burner.

While simultaneously ensuring that Zachary didn't fall out of a tree and Bramble didn't get lost in the woods, we were able to identify hazel, spindle, ash and pedunculate oak in relatively short order. The red twigs of common dogwood were the next to get the botanical once-over, but it was not until I looked a little closer that I realised the shrub's terminal buds were equally as distinctive – with more than a passing resemblance to the cartoon character Scooby-Doo's ears. Taking a break from his rope work in the lower reaches of a nearby tree, Zachary announced himself unaware of the cartoon that had been a staple of my childhood. Declaring that they certainly didn't look like Danger Mouse's ears, he instead suggested they looked like a scorpion's pincers – which I thought was a good shout too.

Taking the opportunity to fill Zachary in on one of my favourite programmes at his age, I mentioned the eponymous dog's endless fascination with food or his 'Scooby snacks', which precipitated a sudden realisation amongst the assembled group that lunch was by now well overdue. A botanical army marches on its stomach, and so after finding a substantial log to use as a woodland pew, Christina unpacked the sandwiches while I set up the stove for what would surely be the highlight of our visit – survivalist's hot chocolate. This would undoubtedly have been the case had the hot chocolate actually been packed, but I belatedly realised that in our rush to get out of the house it had not been placed amongst the other 15kg of random stuff we'd brought along with us. Suffice to say this did not go down well with Zachary, as warm milk on its own was not particularly to his liking. But as the ambient temperature took hold and we quickly cooled down, I'm relieved to report he soon changed his mind.

Sitting quietly, I was suddenly reminded of the presence of birds for the first time that day as the 'cronk' from a pair of ravens flying past punctuated the silence. With birds being my first love, they would normally have been the centre of my attention, but their distinctive call only served to illustrate how focussed on plants I'd quickly become. Ravens tend to nest very early in the new year, and as we caught a brief sight of them through the leafless canopy, it dawned on me that I'd already moved much further over to the green side than perhaps I'd realised.

Reminding myself that I wouldn't find spurge-laurel by gazing upwards, I used the rest of the lunch break as an opportunity to gently appraise my two compatriots of the key features of our remaining quarry, which up to that point had only been conspicuous by its absence. Generally whenever I'm out hunting for a particular plant, bird, mammal or insect (either to film or for my own interest), I've always found the 'searching experience' to be a double-edged sword. On the one hand, it's exciting to use your field skills to home in on your target, with the thrill of the chase helping to build anticipation. However, on the flip side, I've never been able to prevent that feeling of mild rising panic whenever the species I'm looking for doesn't immediately present itself for inspection. The fact that Zachary would be keen for a post-prandial tree-climbing session, added to my doubts that my opportunity was slipping away.

I needn't have worried of course, as Christina cheerfully remarked 'Here it is!' while father and son were busily roping up the next tree for Zachary to shin up. Despite being widely cultivated in gardens, spurge-laurel is also native to the United Kingdom, with its stronghold being in calcareous woodlands. This was one of the few plants in Britain's extensive flora that should be looking its absolute finest in late winter, but Christina's specimen was obviously so young that it was still devoid of its crowning glory: the clusters of small greenish-yellow blooms, which appear at the bases of its uppermost leaves from mid January.

Nevertheless it was still momentous for becoming my first plant 'lifer' (or first sighting in my lifetime) of the project and upon showing the plant to Zachary, I especially enjoyed watching his first impression turn from that of barely feigned disinterest to fascination upon learning that all parts of this glossy-green plant were poisonous. With Zachary having been brought up on a diet of Steve Backshall's *Deadly 60* television series, I knew that any plant that is able to kill is always immediately worthy of a seven-year-old's respect. Even though Christina and Zachary now seemed content that the discovery of the plant had brought about the completion of this particular mission, some people, it would seem, are never happy. So while busily snapping a quick photo of the plant as an aide memoire, my overriding emotions at that moment could perhaps have been described as mixed. While relieved that we'd finally encountered the plant, I still couldn't shake the disappointment that the only specimen we'd found had been such a paltry one.

With the light fading, we then tramped back to the car, with me consoling myself that the trip could have been far worse. I had seen both my target plants, Zachary and Christina had enjoyed themselves and the dog had been well exercised, putting the day firmly into the 'unqualified success' bracket. Re-entering the mud zone, Bramble then brought over yet another stick, in his 373rd blatant attempt to encourage us to take part in the age-old game of throw and fetch. While Christina and I ignored this obvious plea, Zachary dutifully took the stick and threw it down a bank, before quickly following Bramble out of sight.

Slightly worried that the bank led straight and steeply down to the quarry, Christina and I rushed over to see Zachary a few metres below our feet inspecting a plant – not just any plant, but one with glossy green leaves and yellowy-green flowers! They say never work with children and animals, but just occasionally, botanising with them can work out spectacularly in your favour.

One of the few plus sides of the pandemic had been an opportunity to reconnect with my family at a level I hadn't experienced in years. Walks together, board games and movie nights at home became the order of the day, as my normal professional vocation of chasing my own tail became shelved indefinitely. Money worries aside, it had admittedly been wonderful to take my first proper time out in years, but with a strong work ethic firmly lodged in my DNA, by late February I was desperate to get back into the fray.

Despite the comforts and companionship at home, travelling has always been part and parcel of my job, and my feet were becoming decidedly itchier by the week. Plus if I wanted to seriously advance my plant list then I would need to start casting my net a little farther afield than that of short walks or drives from base. So when an opportunity came to film for *The Gadget Show* as their wildlife expert up in Stourbridge, this became too good to refuse.

Keen to hear my take on a variety of cutting-edge gadgets designed to help with recording and identifying wildlife, the producers of the programme were insistent this could only be done in person. As travel at this point was still being heavily restricted, this meant that the production company would need to provide me with a covering letter to use as a 'get out of jail free card' were I to be stopped along the way by Her Majesty's Constabulary. With filming not starting until 2 p.m., my special dispensation to be legitimately out and about had me instantly scheming that it could also potentially be used as a free botanical pass en route.

Early star-of-Bethlehem (Radnor lily, or to use its scientific name *Gagea bohemica*) is a plant of stellar rarity value and one that I'd put into the 'highly desirable but incredibly unlikely' category ever since reading Peter Marren's book *Chasing the Ghost*. The book detailed Peter's attempts to see his last 50 remaining British wildflowers – to varying degrees of success. One of the chapters that had stuck most in my mind was his visit to a site called Stanner Rocks, a couple of hundred metres over the Welsh border and close to the Herefordshire village of Kington. Stanner Rocks National Nature Reserve (NNR) is the only known outpost in Britain for this peculiar and enigmatic member of the lily family, whose main distribution is centred in the Mediterranean. Surviving in shallow pockets of soil in cracks and ledges on a huge south-facing cliff of igneous rock close to the roadside, its continued presence in Britain could perhaps best be described as 'precarious' – both physically and metaphorically.

To make the mission even more difficult, it has been estimated that less than one in a thousand plants ever produces a flower in the United Kingdom, as its preferred reproductive method is seemingly via a series of little bulbils. This, I figured, would make it around a thousand times more difficult to spot, as without the presence of its colourful yellow flowers the only other way of successfully locating the lily would be by finding its narrow, chive-like leaves sprouting out of the rock face. To stack the odds even further against me, Peter Marren was not just in a different botanical league, but had also engaged the help of a local expert who knew both the plant and the reserve well, in what – for Peter – had been a successful mission.

Apart from the worry of looking for a botanical needle in a haystack, the rather more immediate reservation about whether or not I should even attempt a search was the reserve's very obvious danger factor. As it was an ex-quarry, all the online information I could find on the reserve emphasised that visiting the site was the polar opposite of a walk in the park, meaning that a permit would have to be secured before visiting anywhere other than the quarry floor. However, also aware that a plant like this could be just the shot in the arm that my campaign required, I figured it to be a risk worth taking.

Despite my widespread travels I'd never visited Stanner Rocks before and to take in the site on the way to Stourbridge from my home would involve quite a substantial detour. This wouldn't necessarily have been an issue due to my letter of dispensation, but as the reserve entrance was just over the Welsh border, I was also concerned this might be an extra problem. While I had been given permission to travel to Stourbridge, whether this extended to 'via Wales' I was decidedly unsure. After all, Wales had different lockdown rules, and a number of cases had recently made national news when English folk had been caught and fined for crossing the border without good reason.

Trying to put these thoughts to the back of my mind, I left home after an early breakfast, reckoning that with a fair wind I'd have around 90 minutes to look for the plant before needing to depart for Stourbridge. Herefordshire is one of the most sparsely populated and rural of all British counties, and while driving through one small village after the next, I decided it must also have been one of the counties I knew the least too. Justifiably famed for its cider and perry production, the county is also considered the British epicentre for mistletoe, with apple trees being the plant's favourite host. Along with spurge-laurel and early star-of-Bethlehem, mistletoe chooses late winter as its moment to take centre stage, as despite being a hardy perennial, its presence for much of the year is obscured by the host's foliage. But now in winter, picking out the hemiparasite's distinctive clumps from amongst the bare branches proved to be the easiest of botanical ticks as I passed a succession of roadside orchards.

The last stretch of road before arriving at the reserve was along the A44, where the impressive sight of Stanner Rocks suddenly looms into view in the form of a huge, dark and imposing lump running alongside the road. Precambrian by nature, volcanic activity is believed to have been the catalyst for its formation (over 600 million years ago) meaning the reserve contains some of Wales' oldest rocks. Here the dolerite and gabbro igneous rocks weather away very slowly, before eventually breaking down to form base-rich skeletal soils, which supports plants with a calcicolous bent. However, most of these lime-loving plants would not be flowering until much later in the year.

Parking on the other side of the main road, I ventured over the stile and across to the reserve's somewhat dilapidated noticeboard. In addition to a photo of the plant in flower, the sign was dominated by large block capitals warning not just of the site's numerous hazards but also of the flora's fragile state. In essence, going off-piste appeared to be very strongly discouraged. I had read from Peter Marren's previous visit that he and his expert companion had managed to find a single flower on the quarry floor, before then marching up to the top of the crag for a few '*Gageas* with a view'. Without either a guide or a huge amount of time, I figured my best and possibly only chance would be to confine my search to the very same location, or at the very most the cliff's lower ledges, where my binoculars might also be of use in scanning other locations out of reach.

The quarry floor, which was immediately adjacent to the road, was bisected like the net of a tennis court by a post-and-rail fence, which promptly stopped at the point where the cliff face began in earnest. Initially unsure as to which half might be the most productive, I tossed a mental coin and plumped for a scan of the left-hand side. Having left home in bright sunshine, I noticed the day was now looking distinctly overcast with a freshening breeze, making conditions even less ideal for a plant so spectacularly reluctant to flower.

The first blooms spotted along the base of the main cliff belonged to those of lesser celandine. Being such fair-weather flowerers themselves, I took this to be a positive sign that if they were happy enough to put their heads above the parapet then the lily might be encouraged to do likewise. Unfortunately much of the quarry floor was either covered with impenetrable bramble patches or basal rosettes of an array of plants that were still months away from coming into flower, and therefore firmly in the 'impossible for a novice to identify' category.

Sandwiched in between the vertical back wall of the quarry and its flat floor ran a 45-degree slope. The topography here made it far trickier for the brambles to gain a foothold, resulting in larger expanses of bare rock, which were interspaced with sparse patches of vegetation. Trying to put myself in the plant's position, I decided that this subtly different niche was exactly where I'd be hanging out if I liked sun and detested competition.

I calculated that if I were to shimmy further up and across the slope, then a slightly elevated position would not only grant me access to a much more promising location, but also enable me to check out a few more ledges, nooks and crannies that were simply out of sight from the quarry floor. I generally consider myself someone who is broadly comfortable with taking calculated risks, but with the warning words of the interpretive board still ringing in my ears, a moment of doubt crept in as to whether this idea constituted responsible behaviour. While probably not illegal, there was little doubt in my own mind that if I were to be seen by the warden, for example, then it probably wouldn't look too good. And equally if I were to slip and hurt myself then the first phone call made would probably be to the emergency services.

Slyly looking around to see if the coast was clear, which it was, I then quickly made the decision that I'd come too far to give up in such lame fashion and had to give it a go.

Despite being 56, I generally consider myself in better nick than many with '1966' inked in on their birth certificate.

In fact, working my way up the slope proved fairly straightforward, but having somehow contrived to tweak my left Achilles the previous year, I could certainly feel both it and my neck complaining as I carefully shuffled from one uncomfortable position to the next while simultaneously trying to take in my immediate surroundings. Certainly flowers on the rock face were decidedly thin on the ground, but I was nevertheless able to add a few plants to the list as the fleshy belly-button leaves of navelwort and the paired, toothed leaves of wood sage were easily picked out. Three evergreen ferns were also duly spotted from my lofty and slightly precarious position, with the small, glossily triangular fronds and long dark stems of black spleenwort undoubtedly the best of the bunch.

While the visit to Stanner had undoubtedly helped supplement my list with a few plants that I'd simply not have been able to locate closer to home, I'd still not, however, had as much as a sniff of anything that looked even remotely like the numerous pictures of the lily I'd been poring over beforehand.

Retracing my steps carefully back down to the quarry floor, I was not quite ready to give up yet, and – perhaps more pertinently – still had time for one last look. Taking stock for a moment to work out the best course of action, I decided first to remind myself what I was looking for. Thumbing through my BFF, I read that the plant's leaves were 'thread-like and wavy', which also sounded a surprisingly familiar description of the needle-like leaves of fescue grasses. Considering it highly likely that this genus of grass would also occupy the site, I suddenly had this overwhelming feeling that even if I found what looked to be the lily's leaves, how could I be sure? I suddenly felt out of my depth. Instead of embarking on this under-prepared and ill-informed misadventure I could have had a couple of extra hours in bed before gently driving up to Stourbridge ... what had I been thinking?

As I weighed up the pros and cons, the coward in me just wanted to return to the car immediately, while the 'Braveheart' side was already eyeing up the slope again, but this time to the right of the fence line. After a brief battle of wills, I decided to channel my inner Mel Gibson and began once more sashaying up the slope further along. If anything this side wasn't quite as steep, but a big bang to my shin from a protruding ledge during the ascent provided the first painful reminder of the risk I was taking for a plant that might not even be in flower.

Despite almost sliding down a 3m drop at one point, I eventually reached a relatively flat and much safer platform that my ecological instincts instantly told me held more potential. No flowers were immediately obvious around me, but my eye was instantly drawn to a small man-made construction on the shoulder of a small bluff. Consisting of two bricks that held in place a small protective sleeve of chicken-wire fencing, this structure certainly hadn't been viewable from the quarry floor. Realising with rising excitement that the bricks and wire might in fact be a crudely produced attempt by well-meaning botanists to protect a very rare plant from either browsing or trampling, I carefully eased myself into a position from which I could examine what lay beneath.

Unfortunately nothing even remotely thread-like and wavy appeared below. But while this initially appeared disheartening, at least it gave me a glimmer of hope that I might now be looking in the right area. As I searched the surrounding rock-face, a Medusa-like tuft then suddenly revealed itself. Just a shade lighter then British Racing Green, the soft and needle-like leaves certainly didn't appear to have the grassy/sedgy/rushy vibe I'd experienced before – had I just found it?

I decided the only way I'd know for sure was to take a few photos. That way I'd be able to seek verification from someone infinitely more familiar with the plant than me, which might give me some retrospective satisfaction that I had indeed located it. With the leaves digitally recorded I carefully stowed my camera back in its bag before turning round to retrace my steps. As I looked down to ensure I was on a firm footing while not causing undue damage to any fragile vegetation clinging to the rock face, a little yellow flower suddenly pinged into focus, no more than a metre from where I'd been lying prone.

I'd only gone and found it...

As I hadn't expected to find this immense rarity, my first reaction was not one of relief but of genuine euphoria. There it was – nestling in a bed of the very same thread-like leaves I'd just been photographing. I had read that on the few occasions that the plant does produce a flower, a number of the petals often tend to be a touch malformed, and this was indeed the case here too with my specimen – MY specimen! It also looked like one of the other petals had been partly nibbled, but this didn't detract one iota from my find, if anything it made the discovery somehow even more poignant. As I got down to pay homage to this impossibly beautiful flower I also delighted in noticing that each petal had delicate green tramlines, which I guessed would operate as the equivalent of entomological landing lights for the few insects on the wing at this time of year.

Protruding stamens could also be seen, but given that I hadn't seen a single insect all day, pollination seemed more in hope than expectation. Research on the plant's pollen from the tiny colony here at Stanner has also indicated that it may not even be viable, in effect rendering the flower an extravagant waste of the plant's precious resources. It had not been a waste of my resources though, and I for one was incredibly grateful the plant had gone to all that effort. Limping back down to the car, I realised that the grin on my face might last even longer than the bruise I fully expected to soon see appearing on my shin.

CHAPTER 3

The Trouble with Coltsfoot

Needing to record voiceover for a film about birdwatching in lockdown I'd made several weeks previously, I found that a rare visit to Bristol was suddenly on the cards. By now in a state of perpetual anxiety that my plant list was already faltering, I was also desperate to use this trip to the big city for some carbon-conscious botany.

As I'd already discovered on my New Year Plant Hunt when visiting the nearby rural village of Chew Magna, even small conurbations can be surprisingly effective as plant-hunting grounds. And with urban floral diversity surely even higher in cities, Bristol's long history as a trading port might also elevate its plant list to that of few other cities, save London. Furthermore, if I were to get even close to my self-imposed total of a thousand plants, then any worldwide weeds that had successfully settled here would be just as welcome on my list as those plants born and bred in Britain.

Social media undoubtedly polarises opinion, and surely has perhaps as many detractors as adherents. Being one of those who has bemoaned the tentacular grip these digital networks seemingly have on every aspect of modern life, I'm also aware that they can, at times, be incredibly helpful. At their very best, for example, these platforms represent an incredibly powerful tool for gathering together disparate folk with common interests. Twitter has been the only one of these mediums that I'd ever entered into with any gusto, and due perhaps to my television profile have built up a reasonable following of like-minded souls, who appear keen to hear my wild opinions. One of the best facets of this particular platform is its use of hashtags, whereby a certain word or phrase can operate as a rallying call for those with similar interests to come together virtually for the purpose of sharing views and pictures.

With 2021 being my designated 'big botanical year', I'd discovered that many of Twitter's keenest botanists tended to coalesce around the hashtag #wildflowerhour between 7 p.m. and 8 p.m. each Sunday. So by simply following the hashtag I was able to observe in real time what was coming into flower each week. From May to July, when a vast array would be flowering, the hashtag might end up proving less beneficial, but certainly earlier on in the year, it was incredibly useful for reminding me of certain plants' time to shine that would otherwise have slipped under the radar. Each Sunday features a different topic, and with urban flora taking the spotlight on one particular Sunday early in March, a quick trawl of the plants being spotted by others had me entranced and envious in equal measure.

Determined to also see as many of these urban crack-and-crevice specialists as possible, my first job was to draw up a hit list. In the case of my impending urban trip, the plant I most wanted to see was coltsfoot. This familiar and yet patchily distributed member of the daisy family is not just a classic early spring flower but has also been heavily steeped in folklore for the best part of a couple of thousand years. Coltsfoot tea, for example, was often used as a panacea for curing lung complaints and respiratory disorders. The first-century Greek physician Dioscorides recommended it for help with a dry cough, which may well explain why the plant's scientific genus of *Tussilago* emanates from the Latin *tussis ago*, meaning 'I drive away a cough.' As my respiratory system was in perfect working order, however, my interest would be confined to just seeing it rather than using it. Particularly as I couldn't even remember the last time I'd actually seen it flowering.

This lack of personal sightings was certainly in part down to the plant's habit of flowering at a time when I'd normally be paying more attention to birds than plants. Furthermore, as the only charismatic bird that shares coltsfoot's predilection for abandoned brownfield sites was black redstart, I had not given the habitat the full attention that it probably warranted. As derelict and dilapidated sites are often under the constant threat of being redeveloped, this makes brownfield habitat distinctly ephemeral by nature. While on one hand this constantly changing scenario would undoubtedly suit opportunistic species such as coltsfoot, it also potentially meant that the plant's presence at any precise location would only ever be fleeting. So in effect, I wouldn't be able just to turn up at a well-known coltsfoot site, I'd have to find my own.

With coltsfoot, along with many of its urban cadre, not being the kind of plant that is confined to more traditional nature reserves, I'd discovered that perhaps the best way to track down potential locations for further investigation was through the satellite imagery offered by Google Maps®. And it's not until you take this aerial view of Britain's towns and cities that you realise quite how much can be classified as either post-industrial brownfield or just plain derelict. Certainly the centre of Bristol was one such case in point, and while only too aware that some of the sites would inevitably be off-limits to urban botanists, I guessed that there might be more than enough accessible urban decay to bag an interesting plant or two.

I figured that getting my paid work out of the way first, would suddenly open up the whole afternoon for a walking botanical safari. The first location I'd pinpointed online was around Bristol's Create Centre. The building is one of three distinct red-brick and Grade II listed premises that have defined the Cumberland Basin's skyline for over a hundred years. Originally used as a tobacco bond warehouse until the 1960s, the building has since morphed into a gathering point for a collective of environmentally minded companies. The centre is also situated alongside the New Cut, a channel dug between 1804 and 1809 to divert the tidal River Avon through the south and east of the city. It was also the location where I'd previously interviewed the naturalist Brett Westwood some years before, for one of my very few botanically oriented *One Show* films, about how urban plants can help reveal something of the historical development of cities. On that occasion we'd been delighted to find all manner of alien species thriving along and close to the watercourse, ever since their accidental introduction from trading ships when Bristol was a major port for both goods and, to its eternal shame, slaves.

Dispensing with my transport in a municipal car park, I set off on foot to see what could be found. I was already aware that far fewer plants would be flowering than when I'd previously been here with Brett and the television cameras in the summer, nevertheless the species that I initially found were somewhat disappointing. But then, in amongst the usual suspects of lesser celandines, groundsel, smooth sow-thistles and dandelions, I came across a far more interesting member of the carrot family. Alexanders is a fleshy member of the umbellifer family, but is easily distinguished by its distinctive glossy yellowy-green hue. Historically encountered along the coast, it has since ventured further inland, with roadside verges seemingly providing all it needs. Originally from the Mediterranean, it is thought the plant was originally brought here by the Romans in its capacity as an all-purpose spring vegetable. Being partial to a spot of foraging myself, this knowledge would normally have given me a green light to dive in. But given the presence of so much litter and the fact that many of the specimens were comfortably within dog-urinating distance, I decided on this occasion that the plants were perhaps best left right where they were.

The only other species I was able to add to my list here was found growing around the walls of the Create Centre itself. Three-cornered garlic is another native from the Mediterranean region, and was first introduced into cultivation in 1759, and noted as established initially in Guernsey by 1849. Somehow hopping over to the mainland, it then quickly spread along Devon and Cornwall's extensive network of meandering hedgebanks. Previously these locations were the only other places that I'd ever seen it growing in the wild, but no doubt encouraged by our recent run of mild winters, the plant's presence in Bristol was indicative of loftier ambitions. A closer inspection revealed its Toblerone®-shaped stem, leading me to muse that 'three-edged garlic' would have more accurately reflected its most distinctive character. It is also called the white bluebell, and I was pleased to notice on closer inspection – and perhaps for the first time – that a green central tramline ran along the length of each petal. In common with all the other native and introduced alliums, this plant was also positively pungent, albeit not quite as cloying as our native wild garlic, or ramsons. But with the coltsfoot as yet undetected, I took my leave of the garlic, appreciating not for the first time that looking at plants through a different lens was already beginning to prove immensely rewarding.

Walking along the Cumberland Road, my plan had then been to slowly make my way east to the plethora of brownfield sites I'd previously spotted online surrounding Bristol's Temple Meads train station, but this plan was quickly pushed back upon passing a short, cobbled and weedy section of roadside verge. Perhaps no longer than a cricket pitch in length, this section had been elevated from the pavement alongside it, which crucially placed it one step away from where the majority of pedestrians' feet would fall. The undulating nature of the cobbles made it not only more difficult to walk across but also gave the site a small topographical element that the weeds seemed to appreciate.

Upon closer inspection, I suddenly realised that I'd hit the ruderal jackpot. The omnipresent smooth sow-thistle could clearly be seen head and shoulders above the competition, with the more diminutive hairy bittercress equally abundant. But just below the 'tree-layer' in this miniature urban forest was the most wonderful mosaic 'shrub-layer', consisting primarily of common whitlowgrass, with its deeply cleft and tiny white petals, alongside the stickily hairy and reddish-coloured rue-leaved saxifrage.

Never the most gifted of photographers, I decided that this ignored yet much understated floral display simply had to be documented. Suddenly lost in a world of weedy wonderment, I was largely ignored by the various joggers and strollers passing no further than a metre from both the plants and my prostrate form, until a soft, lilting Irish accent snapped me out of my reverie, by kindly enquiring what I was looking at.

'I'm photographing the weeds,' was my obvious but slightly self-conscious response, as I craned my neck upwards to see a masked couple of septuagenarians who were in turn gazing down in bemused interest at the sight of a middle-aged man almost lying in the road. Taking a moment to absorb what I'd said, the lady then replied with the most wonderfully charming response, that 'It wasn't until you took the time to look that you realised how many flowers there were down there.' As my older brother lives close to Galway in the west of the Republic, numerous familial visits had enabled me to acquire a familiarity with a range of Emerald Isle accents. Placing hers as from either Waterford or Wexford, it was her turn to be impressed, with my guess as to her origins being surprisingly close.

Seizing on their obvious interest, I then took it upon myself to tell them about my challenge, while also pointing out a number of the species present. Before bidding me farewell, they both responded in synchrony that the impromptu botanical lesson had made their St Patrick's

Day even more special. Not realising the significance of the day until that very moment, I thought it a shame I'd not been able to show such lovely folk an urban shamrock instead as they headed off down the street. Apart from their kindly and most welcome interruption, the only other person to converse with me for the entire period was a young lad checking on my welfare. Apparently having seen my prone figure in such a location, he'd assumed I'd either blacked out or been hit by a car. After persuading him neither was the case, he too departed, leaving me once more alone with my weeds.

Buoyed by both the plants and the invigorating, impromptu chats, my last bit of business at this location was to identify the one flowering plant with which I was still unfamiliar. Intermediate in height between the weedy trees and shrubs, the presence of numerous four-petalled tiny white flowers immediately helped narrow it down to being another crucifer, or member of the cabbage family. While superficially similar to shepherd's purse, this unfamiliar species was differentiated by the presence of slender cylindrical pods, which contrasted with the familiar heart- or purse-shaped capsules of its far commoner cruciferous cousin. Following a quick consultation with my BFF it became immediately and blindingly obvious that I'd stumbled across thale cress, a species revered by geneticists as the botanical equivalent of the *Drosophila melanogaster* fruit fly, the first creature to have its whole genome genetically sequenced. In possession of a minuscule set of genes, consisting of apparently just 135 mega base pairs (if you must know), this diminutive annual now holds the distinction of being the first plant to have undergone the same level of scrutiny. Perhaps even more excitingly for me, however, it also represented a new species.

This botanical lark was undeniably proving to be quite the buzz, as in just one superficially grotty location I'd managed to bag a trio of completely unexpected plants, with one being a lifer. In fact the only way that this pavement posse could have been even further enhanced was if coltsfoot itself was present. Guessing that brownfield rather than pavement would prove more suitable, I then headed off in the direction of the decay and dereliction I'd previously spotted online – sites that would surely be much more coltsfoot-worthy.

By skulking around the back of Bristol's Temple Meads railway station, my first dawning realisation was how much brownfield acreage cities like Bristol seem to possess. While passing site after abandoned site that had patently seen better days, the second realisation which quickly followed was that virtually every single location was seemingly off-limits to the likes of me. Even the smallest abandoned plot was fenced off, with forbidding fencing adorned by numerous signs warning that trespassers should 'keep out' and these sites were in fact 'private property' that were 'monitored 24 hours a day'. In short, the clear, resounding message was that all these discarded, disused and unloved sites were nevertheless distinctly out of bounds to all but the chosen few with keys to the rusty, padlocked gates.

Beginning to regret not having brought my binoculars along so I might instead be able to botanise from the perimeter, I could only assume the owners of these abandoned sites were either worried that trespassers entering the site might hurt themselves – thereby risking the landlords being sued, or that any intruders might have the audacity to claim squatters' rights. Steadily discounting many of the locations I'd previously found online as being in reality either unsuitable or inaccessible, I then sidled up to a site maybe a couple of hectares in size located just to the north of the train station. Being completely boarded up and screened off by tall panelled fencing on three sides, no doubt to prevent snoopers either looking through or over, it initially didn't look that promising, but the wide open gate on the fourth side revealed a site that at some point had been razed to the ground. Peering through, I saw that it looked superficially like a car park that had long since gone to seed, and in the distance there were some tantalising signs that weeds had moved in.

Moving forward for a better look, I scanned the entrance for 'keep out' signs and to my genuine surprise they were, on this occasion, notable by their absence. Apart from the plants, the only signs of life inside were a couple of guys in hi-vis jackets right down at the far end, who were standing next to a few parked cars and an unmanned digger. Furtively glancing once more left and right, I decided it was too tempting to ignore, before then walking through the gate and onto the site.

Immediately spotting the leafless fleshy scales adorning distinctive purple stems, which were in turn topped off by bright yellow flower heads, I had taken less than 20 seconds to find my prime target sprouting gamely from a crack in the concrete. I involuntarily let out a gasp of delight as I came face to flower with the quirky and yet charismatic coltsfoot. Being such an early flowerer, coltsfoot can often be over by April, but judging by the immaculate state of the bright yellow blooms, St Patrick's Day looked, at least this year, to have been the perfect date. Unlike the groundsel I'd first recorded on the New Year Plant Hunt, the flowers possess both disc and ray florets, and are such a bright splash of sunshine they cannot fail to brighten up even the darkest of March days. Perhaps far more familiar to many are the plant's large polygonal leaves with their white-felt undersides, which appear well after the flowers have shrivelled up and died. For some reason, seeing such a tough little plant colouring up this small forgotten corner of Bristol made me think of the street artist Banksy. Now an internationally famous and famously anonymous Bristol-born urban artist, Banksy's artwork around the city has become retrospectively the source of much civic pride. His work in fact is now so collectable that pieces on the open market can fetch eye-watering sums. But these flowers to me were equally priceless scraps of street art, as they symbolised nature's indomitable ability to reclaim what we've carelessly cast aside.

Extricating my camera to preserve this picture-perfect plant, I was quickly brought back down to earth with a loud 'Oi!' As I looked up from the plant, what must have been a contractor in a filthy hi-vis jacket, with a lopsided and equally dirty white hard hat, was striding purposefully towards me in a way that suggested he was a man on a mission. In a broad Bristolian accent he asked me gruffly what I was doing, to which I was tempted to suggest he mind his own business. But not keen to inflame the situation, I rather tamely, and in an uncharacteristically high-pitched and slightly tremulous voice responded that I was photographing the flowers.

Staring at me like I was the local village idiot he told me in no uncertain terms that this activity was forbidden and I would have to leave immediately. Quickly recovering my composure from this sudden and rude interruption, I realised that I had been so busy gazing at the plant that I hadn't as yet taken any photos and so quickly needed to buy some time before being ejected from the site. Jutting out my jaw, I responded that I wasn't doing any harm and if he'd give me just 10 minutes then I'd photograph the plants and leave. This was patently not the response he'd been expecting, and he reiterated that I'd better leave immediately as the site was dangerous. Sensing the upper hand, I snorted that the solitary digger was unmanned, some 80m away, the site was considerably less dangerous than the pavement just outside the perimeter and furthermore there were no signs forbidding my entry at the gate. Reiterating that I would not leave until I'd photographed the plants, I suggested he should ring his manager, quickly calculating that this would give me a few valuable minutes unmolested with the plants.

Standing less than a couple of metres away, a now-seething contractor followed my advice and was quickly on the phone to a 'Mr Big' elsewhere. Trying to take a few half-decent pictures, however, was proving difficult, while understandably distracted by the conversation I could hear going on in the background. As I lined up the shot I heard the contractor saying that I was refusing to leave, while somewhat more ominously I then caught the voice on the end of the line respond that he was coming down. With my hands by now slightly shaking, which is admittedly not great for macro-photography, I managed to bag a few pictures as a hulking figure passed through the gates, before then lumbering in our direction.

Early thoughts of being able to reason with him went quickly out of the window as I gazed at a face like thunder atop what must have been an 18-stone frame. As I was still on my hands and knees at this point, his six-foot-plus shadow almost entirely covered me as he snarled at me to 'Get… Out… Now!' With Covid social-distancing 'two-metre-plus' rules still very much the order of the day, which he had obviously breached, I gathered all my courage and barked at him to 'Back off!' My testy response was like a red rag to a bull as a volley of accusations poured out of his potty mouth. Shouting at me like a sergeant major would berate an insubordinate private on the parade ground, he yelled I was trespassing on private property and that I'd better leave now or else. While partly fearful that the 'or else' might entail a black eye, or – even worse – a smashed camera, I countered that the site had simply been open and there was no sign forbidding my entry.

Shouting down my reasonable response, the decibel level rose even further as he yelled at me to pack up my stuff and get out, before letting me know how angry he was at having been forced to leave his desk in order to sort me out. But any thoughts of suggesting to him this was probably the first work he'd done all day quickly evaporated when I looked back down at the plants to see he'd crushed the coltsfoot with his size-twelves. While realising the oafish bully-boy hadn't trodden on the plants on purpose, it made me detest this self-important jobsworth even more as I reluctantly packed up my camera bag before being frog-marched off the site.

It was only later in the day, on recounting the whole unsavoury experience to my father-in-law, Graham, that I regretted not having been bolder. My wife's father, a civil engineer of over 30 years' standing, had spent most of his working life on building sites anywhere from Hong Kong to Hinton Blewitt. During that time he'd had to contend with incursions from hundreds of trespassers, and told me that providing I was neither causing criminal damage nor stealing, then the only people legally allowed to remove me were the police. If they'd laid a finger on me, Graham went on to explain, it would have been physical assault and had they touched my equipment I could have claimed for criminal damage. Furthermore, had Mr Big rung the police to eject a flower-spotter, they would in all probability have laughed him off the phone.

Trying to take the positives from my father-in-law's quick tutorial concerning the laws of trespass, I consoled myself that were this to happen again later in the floral season then my argument would be far better prepared. But on just this one occasion, and given the size of the opposition, discretion may well have been the better part of valour. And finally, if it hadn't already been made abundantly clear after my rocky moment halfway up Stanner Cliffs, botany can be a dangerous business!

CHAPTER 4

The Trouble with Covid

As spring steadily and inexorably advanced, the progressively longer and warmer days slowly encouraged more plants to stick their heads above the parapet. My botanical senses were now becoming so finely attuned that as each new plant came into flower, it almost felt like I was either catching up with an old friend or making a new acquaintance.

Covid-enforced lockdown was still pretty stringent though, meaning the only plants I could really add to the list were those either within walking distance or just a short drive from home. With each progressing week I was also becoming painfully aware that while the vast majority of flowers were still to bloom, a small but increasing band – which I had still not yet managed to see – were either currently peaking or already fading fast. While having successfully bagged early star-of-Bethlehem at Stanner Rocks, for example, I had singularly failed to find its more common counterpart, yellow star-of-Bethlehem, at its best-known site in Somerset. It was important during these failures, however, to remind myself that my mission was not to see every *single* species, but somehow I couldn't shake the feeling that each plant missed felt like an opportunity lost.

It is not just individual species that take it in turns to flower as the seasons progress, but different habitats often tend to reach their floristic peak at distinct times of the year too. Heathland, for example, undoubtedly appears at its most splendid in high summer, while saltmarsh plants haven't fully matured until perhaps even later. Flower-wise, the focus during the first half of April tends to move from urban areas across to woodland. But with floral displays across forest floors decidedly ephemeral by nature, this was one botanical bus I could not afford to miss.

I often describe the Chew Valley where we live as pastoral countryside at its best. The landscape is low-lying, and yet undulating, with compact and affluent villages surrounded by farmed land. Many of these villages date back to the time of the Domesday Book, with evidence of humans taking advantage of the valley's fertile clay soils as far back as the Stone Age. Given the value of this farmed land through the millennia, the end result of this relentless drive for agricultural activity has resulted in semi-natural or ancient woodland being a surprisingly rare commodity within the Chew Valley.

Keen to up my quota of woodland plants during this optimal time, I'd certainly received some help, courtesy of the numerous hedgerows close to home, which had delivered a variety of woodland-edge species, such as cowslip, various violet species and greater stitchwort. But the 'proper' woodland plants would once again necessitate travelling a little further afield.

To help my cause, a timely and further relaxation of Covid regulations announced on 29th March also meant that up to six people (or two households) were now permitted to meet up outside. I'm very fortunate that my in-laws not only happen to live close by, but also (and fortuitously) inhabit a house surrounded by woods just to the east of Bath. Perhaps a well-overdue visit might allow us to kill two birds with one stone.

Aware of my mission, my father-in-law was not only eager to see us but was also keen to help me by proactively taking us to a scarcely visited part of the wood where wild daffodils grew. Surely one of the most cultivated of all our native plants, the spectacle of wild daffodils growing naturally in an English woodland is a sight that has been on the wane for years. Not much of a gardener myself, as I always prefer the real thing, to my purist botanical mind the sight of the wild species in its natural setting will always be infinitely more appealing than the serried ranks of gaudy cultivated daffs that now seem to clutter almost every roadside verge in early spring.

Upon meeting up just outside Graham and Laura's cottage, the ensuing walk in the woods was a success on all levels. Not only did it enable Christina to have a much-needed catch-up with her parents and Zachary with his grandparents, but as we strolled through the trees with a carefree feeling that had been markedly absent for the previous 12 months, I was also able to pick up a few key woodland species to boot. Wood anemone, wood sorrel and opposite-leaved golden saxifrage were all spotted before the drifts of daffodils that Graham had promised us finally came into view. Perhaps middling in size when compared with the countless garden varieties, the creamy-yellow petals and darker yellow trumpet of the wild species somehow managed to relegate all the cultivated forms to little more than pale imitations. In my heavily biased opinion, the original was still far and away the best.

Blissfully happy with such a fine display of my target species, I was unaware, however, that the daffodil spectacle would prove to be most wonderfully upstaged on our walk back. Toothwort is such a memorable plant that I can remember with utmost clarity the previous three occasions when I've been lucky enough to chance upon this pastel-pink parasite. The last time must have been around 15 years earlier, in my golf-playing days, when a wayward tee shot into the undergrowth led me to a clump discretely tucked away at the base of some hazel scrub.

On this occasion, however, it had been Christina who had spotted around four or five spikes, once more associating with hazel. After giving my wife a congratulatory hug, I assumed my now-default position of kneeling down in front to both pay homage to, and for a better look at, this mercurial and ghostly plant. While certainly being one of our most distinctive plants, toothwort could never be described as one of our prettiest. Superficially resembling an orchid, on closer inspection it reveals a jumbled column of flowers that almost appear like a stack of dirty, mauve-stained molars – and from which it derives its name.

With toothwort unable to produce its own sustenance due to a complete absence of chlorophyll, it has evolved to parasitise a number of host trees, with hazel considered the most commonly targeted species. Toothwort achieves this daylight robbery by first latching onto its chosen host's roots before then siphoning off all the nutriment it is ever likely to need. It's perhaps fitting that toothwort has also been called the 'corpse flower', but the origin of this name might equally come from the plant's fleshy tones, which supposedly resemble the pallid colouring of someone who has just died. Certainly toothwort's frequent presence in graveyards further enhances its reputation as one of life's takers rather than givers, but on this occasion it had certainly provided me with a gift – a species I hadn't expected to be given, but was certainly glad to receive.

Eager for one last shot at securing a few more woodland flowers before they largely disappeared, my online research led to only two other possible candidates close to home: the Avon Wildlife Trust's Folly Farm and a small, isolated woodland called Bithams Woods, just north-west of where we live.

Folly Farm's name is somewhat confusing, as its principal habitat consists of a reasonably large chunk of semi-natural woodland with a fine associated flora, but the fact that dogs aren't allowed there made the reserve a site we rarely visited anyway. By contrast Bithams Woods' remoteness meant it was rarely visited by anyone at all. With the Easter school holidays still upon us, I was also keen to use this visit as another opportunity to encourage Zachary's participation in the Great Plant Hunt. And while weighing up the pros and cons of each site, what eventually swung it Bithams' way had been the discovery of a floral list that included a plant that had me salivating in anticipation.

Moschatel, or town-hall-clock, is an early flowering ancient woodland indicator, which, due to a combination of limitations on travel and the lack of quality woodland nearby, I'd realistically given up any hope of adding to my list. Confined to woods, hedgebanks and occasionally mountains, this peculiar little plant is the only British member of the obscure *Adoxaceae* (or Moschatel) family, and a plant that I'd previously only seen twice before. Generally looking its best in early April, the plant's small yellowish-green flowers are arranged with its four 'faces' at right angles to each other – in the manner of a town hall clock. These 'clocks' then rise on floral spikes above a bed of trefoil leaves, and where the plant is present, it can often form quite extensive patches.

As Bitham's Woods was just a short drive from home, we'd made a family visit to the reserve a few years previously, with my one clear memory being how little of the site was accessible from established paths. As Christina was by now back at work, the hastily convened and somewhat depleted Chew Stoke Botanical Group on this occasion would comprise myself, Zachary and Bramble – meaning any plant-spotting would have to be opportunistic rather than systematic. The reserve itself can only be reached by following a little-used public footpath across a couple of large fields, before it then cuts just inside the reserve's westerly perimeter. With no other woodland even remotely close by, the reserve is effectively an island surrounded by a sea of improved pasture fields. In fact on the Ordnance Survey map, the shape of the 6-hectare site appears remarkably similar to that of a sea turtle, complete with head, flippers, hind feet and carapace. Here the turtle fancifully appears to be swimming to the nearest water source, which in this case is Chew Valley Reservoir, a couple of kilometres away to the south-east.

As Bear Grylls was still proving to be of profound influence in my son's life, Zachary's participation on the trip was sold on the promise that tree-climbing would figure every bit as prominently as flower-spotting. And after deciding that the one solitary path at the hind feet end of the reserve offered little in the way of adventure, the intrepid trio opted instead to go off the beaten track by plunging directly into the woodland.

There is something invigorating about being in a place that has so few visitors that there is hardly any evidence of human activity, and as we wandered between the towering oak and ash trees it almost felt as if this small, secluded block of woodland was our own personal fiefdom. Normally I would be tentative about spending too much time in the middle of a woodland with a large proportion of ash, as the dieback currently decimating this species can result in the sudden and unpredictable shedding of limbs in any trees falling foul to this disease. However, the ash here looked in remarkably good nick, with the wood's isolated nature clearly having helped to protect it from the worst effects of this highly pathogenic fungus.

Dead wood is of course an essential habitat in its own right, being the primary source material for countless different invertebrates and fungi, and judging by the fact that Bramble was spoilt for choice with sticks littering the woodland floor, this was definitely a reserve that had avoided the plague of being 'over-tidied'. Managing woodlands for wildlife can be a delicate balancing act, as some maintenance is vital for the habitat's long-term health. A chainsaw, for example, is often the most important conservation tool at a forest ranger's disposal.

Reducing the number of native trees (or thinning) will often be to the betterment of those left behind, while the removal of non-native species should be equally beneficial across many of our deciduous woodlands, which, just like our brownfield sites, have often become neglected. It's also widely agreed that there has been insufficient coppicing across most of Britain's woods for decades. Adopted initially as a method for 'farming' timber both for fuel and as a building material throughout the ages, coppicing is steadily becoming reinstated as its importance for preserving biodiversity has become fully understood. By felling certain species of trees at their base (or stool), before then allowing a number of stems to regrow in place of the original single trunk, coppicing also allows more light to penetrate down to the forest floor – with the ground flora benefitting almost immediately.

Say it quietly, but woodland interiors at any time of year can often be a bit of a wildlife-free zone, with research showing that a far greater number of species inhabit the first 10m of any wooded perimeter than live in the entire remainder of the wood itself. Certainly anyone keen on birdwatching will tell you that forest borders are always busier for birds, as along these edges you may well also encounter birds of the wider countryside, such as those preferring scrub or farmland, in addition to the woodland specialists. The same can also be true of plants, with relatively few species thriving in woodlands' shady centres. The exception to this, however, will be in those parts of the interior close to woodland glades and rides. Here, the more open features perfectly mimic those of the woodland's edge, which will in turn attract the types of plants that need a little more sunshine in their lives.

And this is where there was clear water between mine and Zachary's interests – as the woodland's centre offered the best trees to climb, while Moschatel would undoubtedly be better suited to somewhere less shaded. Of course, being a parent is all about taking one for the team, so the next hour was duly spent casting a watchful eye over Zachary's above-ground exploits in the dark and dank interior, instead of along the woodland's sunnier fringes.

Eventually we did manage to cover most of the woodland, albeit in different ways, with Bramble and I completing it on foot, while Zachary took the arboreal route. Having previously watched *Tarzan* on television, he was particularly taken with how Edgar Rice Burroughs' fictional character had frequently used lianas to get around the African jungle – with the closest vegetative equivalent at Bitham belonging to the woody stems of honeysuckle. Testing the weight-bearing capacity of each sturdy stem in turn, and aided by his father's diligent attention, he eventually swung his way in anticlockwise fashion round to the very north-west tip of the wood and back onto the one clearly defined path we had initially eschewed. While Zachary and Bramble had been having a ball, I'd drawn a complete blank on the Moschatel, without as much as a single plant added to my now-stagnating list.

But as I quickly surveyed the footpath's position along the woodland's edge, the increased light levels told me all might not yet be lost. The floral diversity was higher here than anywhere else I'd seen in the wood, with the 100m stretch back to where we'd initially entered clearly offering my last and best shot at redemption. Having also been the model parent so far, I could now enjoy 'me time' as my attention switched to the path-side herbage. Dog and child suddenly became a distant memory as the green mist descended. That was until my concentration was suddenly punctured by Zachary yelling, 'Daddy, look at me!' Reluctantly tearing my gaze from the woodland's floor, I could see him 15m further along the path and around 1.5m off the ground as he hung onto the underside of a particularly sturdy honeysuckle stem, much like a sloth would in the branches of a cecropia tree in Costa Rica.

Snap! Unable to cope with the weight of a seven-year-old, the stem suddenly and spectacularly broke and I watched Zachary plummet like a stone before hitting the ground with a thud. Sprinting over, I found to my immense relief that no bones appeared to be broken, and judging by the gasps, in between the tears, he had just been winded. I found a nearby tree stump for him to sit on my knee while waiting to catch his breath; it quickly returned with a wail as the woodland suddenly became somewhat less tranquil than just before the incident. Fortunately a chocolate bar can help with both the physical pain and the alleviation of embarrassment and after a few minutes he was right as rain. Looking down at his feet, perhaps more in contrition than anything else, he stared at the ground, before then looking back at me with a beatific look on his face. 'Daddy, is this what you're looking for?'

Realising instantly that he'd seen something different, we immediately dropped down to plant level, before then looking on in wonderment at the 30 or 40 Moschatels, or 'Little Bens' lining the path ahead of us.

During the planning stages of my big botanical year, I decided that it would be important to make the trips as carbon-conscious as possible. An environmentally aware and successful campaign would involve first covering sites close to home, while trips further afield would, where possible, be either tacked onto the beginning or end of paid work elsewhere – much like I'd managed at Stanner Rocks, for example. But there would need to be a few plants where I'd have to make an exception.

Sand crocus is nothing short of a botanical rockstar. This diminutive member of the crocus family used to belong to an elite group of plants so rare that they have either been reduced to, or only ever recorded from, a single site in Britain. From 1981, for example, the sand crocus could only be seen at Dawlish Warren in South Devon. However, in May 2002 it was relegated from the Premier Plant League when rediscovered near the village of Polruan in southern Cornwall after a gap of precisely 21 years. The crocus's presence at this second site, however, in no way diminished its cachet in my own mind, and was a plant I simply had to see.

Late March to mid April is generally considered the best time to see the crocus's six pale purple petals gently unfurling in the weak and watery spring sunshine, and as part of this window also coincided with the second week of Zachary's Easter holiday this presented another golden opportunity to double up family time and plant-spotting time. Like all parents, Christina and I were constantly casting around for ways to keep our boy entertained, but getting my wife to agree to another wild botanical excursion had to be handled delicately.

Carefully picking a moment, I casually suggested that as we hadn't been to the seaside for well over a year, perhaps a day trip might be just the ticket. Knowing me too well, she immediately saw through my flimsy façade, and with narrowing, suspicious eyes demanded to know why I really wanted to go. Deciding that honesty was now my only option, I confessed that I would love the opportunity to see an exceptionally rare plant called sand crocus. Watching her anxiously as she mulled over the idea, I used this pause for thought to lay on thickly how beautiful the Exe Estuary was, and with the guaranteed promise of both sand between our toes and ice creams on the beach, what was not to like?

To be honest, it was an easy sell, and with Christina now on board I had already won half the battle. The only other issue before the buckets, spades and eye lenses could be packed was whether our trip breached Covid's regulations. The 'stay at home' rule had ended on 29th March, but the government still advised that working from home should continue where possible, and that travel should be minimised.

Were we to be stopped, I reasoned, I could justifiably claim that 'twitching the sand crocus' was not only part of my paid work, but also something I patently couldn't do from my office window. Despite this, a recent news story of the discovery of a northern mockingbird in an Exmouth garden – and just a few miles from the crocus's location – had me more than a touch worried. Only the third record for Britain, this immense avian rarity, which is usually confined to North and Central America, had been spotted in a suburban Devon garden just a few weeks previously. And in the ensuing mass twitch, five birders had been caught and fined under the new

Covid legislation for driving halfway across the country. In the intervening six weeks the rules had been relaxed somewhat, but was my plan any different to what they'd been fined for? There would be only one way to find out.

As we set off early in the direction of the M5, the cold and blustery conditions weren't ideal for tracking down a decidedly fair-weather plant, but with commitments on the other remaining days before Zachary returned to school, it was now or never.

Upgraded to a National Nature Reserve in 2000, Dawlish Warren is of such importance for wildlife that it has every possible designation under the sun. A Site of Special Scientific Interest (SSSI) since 1981, the site is also a Special Protection Area (SPA) and a Special Area of Conservation (SAC) – in one word the reserve can be perfectly summarised as 'special'. The sand crocus has made its stand amongst the Warren's dune slacks – the low-lying and occasionally seasonally flooded areas situated between the ranks of undulating dunes running parallel to the beach. A large proportion of the sandy peninsula now comprises the Warren Golf Club, with my research indicating that most of the plants appeared to prefer the tee boxes of the first and second holes, while a lesser number can also be seen just behind the visitor centre. Apparently this smaller population had been translocated from an area close by that had been sacrificed to make way for a car park. This meant these hand-sown plants would not have quite the stellar provenance of those growing across the golf course, but keen not to be caught trespassing again so soon, meant the 'lesser crocuses' – on this occasion – would be more than adequate.

As we arrived at the aforementioned car park, the first thing to greet us upon jumping out of the car to obtain our bearings and stretch our legs was a cold north-easterly wind. Christina had wisely realised that with the beach beckoning just beyond the dunes, the proposition of sand instead of a boring old dune slack would be infinitely preferable for boy and dog. And with

Zachary's tree fall still fresh in both our minds, she kindly said she'd leave me to it for an hour or so. Much as I loved my family, this freedom to focus on the crocus was music to my ears. Once the botanical rarity had been bagged, this would see the pressure off and ultimately mean the family could then be reunited for an enjoyable and stress-free day at the beach.

Bar a few passing dog walkers, there was no one around at the nearby visitor centre, which was not only shuttered up but also looked like it had seen slightly better days. The all-weather information boards outside briefly mentioned the presence of the crocus, but most of the wildlife particulars concentrated on the birds to be seen on the estuary behind me, leaving me to lament once again that when compared to birds, fabulously charismatic plants always tend to be relegated to little more than an afterthought.

The sandy grassland just behind the centre, and the area where the crocus supposedly resided, was laid out before me like a gently undulating bowling green. With the grass no longer than you would find at Wimbledon's Centre Court, it was obvious that this closely-cropped turf was just what the sand crocus required. Being a plant that detests any competition, grazing is considered crucial to its survival, and without the presence of a large rabbit population the plant would have long since departed from here too. Presumably on the golf course, the mower would have done effectively the same job as the rabbits, I mused, while assuming my now-familiar plant-hunting technique of stooping down, with my neck craned and feet carefully placed as I slowly began to stalk my prey.

As it was still so early in the year, the vegetation could hardly be described as blooming, but under my watchful gaze, small, discreet flowers soon began to reveal themselves. A minuscule *Myosotis* was the first plant to get the treatment, which proved to be early forget-me-not, a plant that if I'd probably seen before, I'd certainly never successfully identified – until now. My run of early-season crucifers then continued with shepherd's cress, a very localised plant of sparsely grassy or sandy places, which was growing in abundance on the slightly raised areas. Each specimen was no taller than a 15cm ruler, with almost every single flower shut tight, and given the biting wind who'd blame them? But in a couple of flowers I could just make out the unequally-sized petals that are such a prominent feature of this decidedly uncommon crucifer. And there was certainly no mistaking the plant's dainty, rounded and distinctively notched seed-pods. Next to get the treatment was one of the two parsley pierts. But with these two closely-related species difficult to split even for talented botanists, my valuable time would be far better spent taking a few close-up shots, before then consulting Twitter at the next #wildflowerhour.

With the botany both absorbing and exhilarating, I had still yet to see any sign of the sand crocus at all. Taking a moment to remind myself as to what the main target looked like, I consoled myself that even if the flowers were closed up then surely I'd be able to spot the crocus's tentacular, thread-like leaves ... but was I even searching in the right place? Looking around, I was unsure where to hunt next, and the only other folk I could see were a couple of guys repairing the fence next to a path over the dunes, and along which my family had passed an hour before. Deciding to find out if they knew anything about the plants, it seemed like I was in luck when one declared himself to be the site's warden. He confirmed that I had been indeed looking in pretty much the right location and gestured towards a spot immediately in front of a couple having lunch, who had entered unnoticed while I'd walked away. Retracing my steps, I began an even more detailed search closer to the couple, but not wanting to crowd their personal space, and for fear of looking slightly odd, I kept a respectable distance. All I could find on this second search, however, was more shepherd's cress. And with the couple appearing not even remotely interested in the plants, I figured asking them would be a waste of time.

Getting more frustrated with every passing minute, and to add to my now-rising sense of urgency, I also had a sinking feeling that my phone would soon be ringing to demand my presence on the beach. Now within earshot of the lunch couple, I hadn't meant to eavesdrop but suddenly heard the man mention parsley piert – surely only 'proper' botanists would mention such an obscure plant! Aware that I had little time to lose, I butted straight into their private conversation by asking if they'd seen the crocus, to which the lady responded in panto-esque fashion with 'It's behind you!'

Seeing my confused face she then elaborated by saying the crocuses only seemed to be present in the very shallow hollows, which may well have stayed wetter for slightly longer during the winter. By now, I had lost a little faith in my ability to ever find this plant, so the lady, at my request, took me some 10m back towards the visitor centre, much like an adult would take a toddler to the toilet, before quickly pointing out one in flower. My eye suddenly felt like a camera lens being brought sharply into focus – the plants were everywhere!

Due to the inclement weather, few had bothered to open their single flowers, but the curly green bristle-like leaves, which brought to mind Medusa's snake-locked hair, could now be seen carpeting a couple of the hollows. Despite being undeniably dinky and with only a few flowers open, I was nevertheless bemused as to how I'd missed them. Thanking the couple profusely, I then lay along the turf before taking a few pictures of one particular specimen that was flowering beautifully. Each perfectly formed petal had a distinctive dark purple vein running from tip to base, which all pointed towards the bright yellow and pollen-plastered anthers. My reverie was then interrupted by a buzzing in my pocket, indicating my time was up. 'Come and see what I've found, I somewhat brazenly said, as I allowed my mind to drift forwards to a shoeless walk on the beach complete with an ice cream – and without a single Covid policeman in sight.

CHAPTER 5

Just Beyond My Doorstep

While the immediate vicinity of the Chew Valley was of minimal botanical interest for anything other than the relatively common and widespread plants, we were fortunate enough to be sandwiched in between two outstanding botanical areas in the form of the Mendips to the south and the Avon Gorge to the north. With both categorised as Important Plant Areas (IPAs) by the wildflower charity Plantlife, each also contained enough rare plants to stimulate the salivary glands of even the most seasoned plant hunter.

The only problem here, certainly from a botanical point of view, was that I hardly knew either site at all. Despite having spent my entire first year after moving down to the West Country living a mere stone's throw from the iconic Clifton Suspension Bridge, my knowledge of the precise locations of the Avon Gorge's rare flowers was (somewhat embarrassingly) little more than an educated guess. Equally, with 10 years under our belt in the rural idyl of the Chew Valley, a working familiarity with the numerous footpaths criss-crossing the Mendip hills close by had given me surprisingly little insight as to where I might start tracking down its many botanical treasures. In essence I needed help, and fortunately this was at hand following a quick internet search that included the words 'rare', 'plants' and 'Somerset'.

The Somerset Rare Plants Group (SRPG) consists of a merry band of Somerset's finest botanists, whose goal is to promote the study and interest of all the vascular plants growing wild within the county's boundaries. Set up in 1997, the aim of the group was to gather together all the botanical expertise in order to maintain the long and fine tradition of recording wild plants within the county. Even more helpfully for my cause, they were always keen to sign up new blood. So after a quick email to the membership secretary, and the payment of the smallest subscription going of any members' group I'd ever known (£8), I had become a fully signed-up member before you could say 'Cheddar pink'.

In a system that was devised by botanist H. C. Watson in the 1870s for the purposes of botanical surveying, the whole of Britain is broken up into a series of Watsonian vice-counties to create 113 units of roughly equal size. The now geographically defunct county of Somerset, for example, was split into South Somerset (labelled VC5) and North Somerset (VC6). But despite the botanical Mecca of Avon Gorge being just across the border in West Gloucestershire (VC34), I was assured by my email contact, and now new best friend, Clive Lovatt from the SRPG, that he would certainly be able to point me in the right direction. In addition to responding to requests from annoying people like me, the group also carries out regular field meetings to interesting sites within the county, but under the spectre of Covid these had been shelved for the entirety of 2020. However, with most trips outside and of course in the fresh air, there was considerable optimism amongst the group that these would recommence at some point in 2021.

Provided the relaxation of lockdown rules continued along the same trajectory, these excursions might prove to be a lifeline later in the year, but in the short term I needed help in tracking down two of the Avon Gorge's most illustrious plants. Bristol rock-cress and honewort both possess a rich history that, down the centuries, has seen them draw some of Britain's most celebrated botanists to the iconic carboniferous limestone cliffs towering above the River Avon. And I was keen to add my name to that revered list too.

Labelled as 'the original rare plant locality' by the author Peter Marren in his seminal work *Britain's Rare Flowers*, the Avon Gorge was the site where honewort was first discovered by William Turner, the author of the first English herbal (a guidebook for the medicinal use of plants) as far back as 1562. This early discovery makes it one of the first rare plants to be documented in Britain. Found across southern Europe, this unusual umbellifer is limited to just a few other sites in Somerset and one location in Devon, with the Gorge being the most northerly place in which it grows.

Bristol rock-cress, however, was not found in the Gorge until reported to the famous botanist John Ray by his colleague James Newton in 1686. It was first found on St Vincent's Rock, which is now perhaps better known as the foundation for the Clifton Suspension Bridge on the eastern (or Bristol) side. The rock-cress also possesses a different level of rarity to honewort, as within the United Kingdom it is known solely from the Gorge, while also being found at a few isolated sites in the French Alps and the Pyrenees.

As anyone who has visited Bristol before will agree, the Gorge is as huge as it is impressive and as I wasn't endowed with the botanical skills, patience or time that Turner and Ray possessed centuries before, Clive's help was essential. So after a few emails back and forth, which gave me the opportunity to explain both my mission and honourable intentions, he generously passed on both an OS grid reference for the rock-cress and a location in which to search for the honewort. The Easter holidays were still in full swing, but given the potentially hazardous nature of searching for rare plants halfway up steep cliffs with an accident-prone and recently-turned eight-year-old, Christina and I decided that this endeavour would have to be a lone mission.

Clive had told me that the rock-cress was dotted at a number of sites around the Gorge, but as many of the precise locations were almost impossible to reach without climbing equipment, he'd pointed me towards perhaps the most accessible spot, which was just beyond the Clifton Observatory. Parking close to where I used to live, it was immediately obvious from the number of folk milling around that many others had also decided to take advantage of the year's first genuinely sunny day to enjoy the splendid views of the Gorge from the bridge. Armed with my grid reference, I walked up to and past the observatory with a growing sense of optimism that the final loosening of the Covid shackles would provide me with the impetus to kick on with my ambitious campaign.

An eight-figure grid reference is an almighty leg-up when trying to track down a small plant in the middle of a metaphorical haystack, as it basically reduces your search area to nothing more than a 10m by 10m square. Clive had also passed on the invaluable information that the plants were to be found a few feet up on a rocky outcrop which ran alongside a sunken path, and upon reaching what appeared to be the location from his description I spotted a small section of limestone jutting out of the closely-cropped turf. Immediately behind this spot was a park bench, where a couple of lads were busily working their way through a four-pack of cider each. My feelings of self-consciousness increased as I squatted down just a few metres in front of them to inspect what might be growing around the rock. The first flower to be identified here was one I'd seen many times before, that of creeping cinquefoil, but apart from this nothing else stood out as even remotely interesting, or in flower.

Suddenly I heard a broad Birmingham accent, as one of the bench lads – obviously unable to curb his curiosity any longer – shouted to ask what I was looking for. Only too aware that 'flower-spotting', to those that have not seen the light, might initially seem like a prime candidate for ridicule, I was initially reluctant to engage. But – proving you should never judge a book by its cover – upon telling them about my hunt for a rare plant, I could not have found them any more courteous and receptive. Kane, the Brummie and Justin from Plymouth both worked as steel erectors, it transpired, a job that I was to learn involved erecting skeleton frameworks for construction projects. Having knocked off early for the day, both were keen to enjoy a few pints in the sunshine before returning to their digs close by. Enthused by the fact that both had declared a passion for wildlife, I showed them a picture of the rock-cress, before then proceeding to talk about the bigger project that would dominate my life for the next few months.

Justin then mused out loud that the rock-cress must be incredibly rare to be named after a city, and this casual but profound off-the-cuff remark suddenly made me realise that many of our rarest plants do indeed have a location attached to their common names, like Breckland speedwell or Teesdale sandwort – why hadn't I thought of that before? Delightful though the chat was, it wasn't helping me find the plant, so after wishing them and their erecting well I moved slightly further down the path. Quickly reaching a point where the path had been diverted around a rock, I instantly realised this fitted Clive's written description far better than the spot where I'd previously been looking and quickly found the rock-cress within a matter of seconds.

'It's only taken twenty-one years to get round to seeing you,' was my first reaction, as I took a moment for the delight and relief to sink in. Despite the fine weather, the rock's east-facing orientation meant the few plants present were now firmly cast in shadow, with the result that each flower's four creamy-white petals were in the process of slowly shutting up shop for the day. Easily 15cm in height, the plants were taller than I'd imagined, with each bloom's slender and almost leafless stems rising like a miniature green phoenix from a dark-green splayed rosette of basal leaves. The soil to which these rosettes were anchored must have been little more than skeletal, making their hold appear precarious to say the least, and given their close proximity to such a busy thoroughfare it would only have taken one carelessly-placed boot to mash them into the rock. Their continued presence here was nothing short of a miracle.

Before getting out my camera I decided that the plant needed a more appreciative audience and so trotted back to my new construction friends, who appeared only too pleased to take a quick break from their cider consumption to catch up with a botanical mega. 'That is the rarest plant you'll ever see,' I stated with conviction to the delightful duo, who then proceeded to take a few phone pics of Bristol's pride and joy to show to their landlady later on. Bidding farewell for a second time, I then got down to the serious business of taking my own photos, while musing that sharing rare plants with other enlightened folk is almost as enjoyable as finding them in the first place.

The Circular Road, as its name suggests, is a winding route around the west of Bristol's famed Clifton Downs, which ultimately leads to the steepest sections of cliff on the Gorge's eastern side. But before this well-known viewpoint is a far less commonly visited section called The Gully, where the slope's more gentle nature even allows for the presence of one of only two walkable routes which directly link the Downs above to the Portway below. Accessed by a recessed gate set back from the road, this clearing through the trees leads to a flat grassy terrace, which is itself separated from the top of the gully by a large chain-link fence. Possessing gaps at either end, this fence had patently not been installed to keep people out of the gully, but to prevent anyone from accidentally taking a tumble down the rock face.

The enclosed nature of the flat terrace not only gives it a 'secret garden' vibe, but its elevated position also offers up fine views across the Gorge. To add even further to its attractiveness, a west-facing aspect turns into a real sun-trap on warm afternoons, but upon passing through the gate, it wasn't the extra warmth or tremendous vista that came to my attention first, but the unmistakably sweet, cloying pong of marijuana and the sound of reggae. The gully was obviously not just a site of national botanical importance but also the perfect party venue.

Avoiding the temptation to join in with the terrace revellers, I headed straight for the fence before then ducking round for my first view of the rock face. While certainly less precipitous than I'd previously experienced at Stanner Rocks, the slope would still command respect, I reflected, as I carefully zig-zagged my way down.

Clive had warned me that where I placed my feet while traversing such a botanically sensitive area should not be my only worry, as I would also need to keep an eye on what was happening above. And upon spotting the first of many broken beer bottles and cans that had been lobbed from above, I suddenly realised what he had meant. Way down below me the A4 and the River Avon beyond were clearly visible, before I then picked out a herd of goats navigating their way around a more precipitous section on the gully's other side. For many of our rarest plants, the biggest threat to their continued survival, apart from the existential climate crisis, comes from being scrubbed over by the likes of bramble and gorse. Much like the sand crocus, which desperately needed help from the rabbits and mowers, Bristol's rock-loving rarities also required help keeping the competition in check. Here, the goats' abilities to reach virtually any corner and eat even the most unpalatable of shrubs had made them the consummate management tool for what was a tricky habitat to conserve.

They say 'Cast no clout till May be out,' but due to the exertion of descending and the sunny nature of the afternoon I soon found myself down to just a T-shirt. Plants I'd already seen growing along Bristol's roadside verges were quickly spotted, as rue-leaved saxifrage and common whitlowgrass jostled for position on the small, limey ledges alongside another plant with which I had far more familiarity. Common rockrose must surely be one of the signature plants of many calcareous grasslands. Often appearing like a creeping shrub, it possesses prostrate woody stems, from which sprout wiry offshoots covered in dark green strap-like leaves. But by far the most distinctive feature of this delightful plant are its five sulphur-yellow petals, which always seem to have the appearance of gently crinkled tissue paper. Common rockrose is generally a plant that only flowers from late May onwards, but obviously a combination of the sunny weather and the site's aspect had encouraged a few blooms in one discreet patch to open as much as a month earlier than would have normally been expected. Also seemingly flowering earlier than its allotted slot was another lime-loving plant in the form of salad burnet, which was easily identified by its small globular flower heads that sit atop a rosette of red stems, each adorned with pairs of toothed and rounded leaflets. As its common name suggests, the plant's vegetative parts are considered an excellent addition to salads, and unable to resist a quick nibble, I enjoyed its gentle cucumber-like perfume while continuing my search of the rock face.

Tucked away in a few discreet locations were also the first tantalising signs of a plant I hoped to return for later in the year. Round-headed leek (or Bristol onion) is another Bristol speciality, which, like the rock-cress, is confined to the Avon Gorge. Identified as a 'native', there is however some doubt as to the plant's real provenance, because despite its showy and distinctive flowers resembling long, purple lollipops, it was not discovered until as late as 1847. Some 161 years after the far less obtrusive Bristol rock-cress, for example. In fact the Gorge's bare white limestone slabs are also home to quite the variety of alien alliums. With plants such as rosy garlic, honey garlic, field garlic and few-flowered garlic having also become naturalised on the cliffs here, the question has long been asked as to why the status of the Bristol onion should be elevated to that of native, while the other alliums have been relegated to mere introductions.

Determining that 'the onion' could only be added to the list when flowering in high summer, I renewed my focus on my primary target and then suddenly right next to a smashed brown beer bottle there it was ... or there I think it was. My first reaction was 'honewort!' only then for the doubts to start creeping in. It hadn't quite started to flower yet, but the pinky-white blooms were quite discernibly in distinct umbels – a key feature of any member of the umbellifer family. However, most other umbellifers, like cow parsley for example, are reasonably tall and leggy, while this specimen was wedged into a crevice, almost like a tick. The stems and leaves were solid with a distinctly fleshy appearance, making it look almost like a succulent, which was a feature that my BFF guidebook hadn't highlighted. As I racked my brains, the only other plant I could think of that looked even remotely like this was rock samphire – but did that even grow here?

Honewort is a plant with a fascinating lifestyle, which, like holly, has separate male and female plants. Pollinated by ants, honewort can take up to 16 years to flower, but this magnificent crescendo also forms its swansong, as once its genes have been encapsulated into fruits then the plant will simply shrivel up and die. If this were indeed the plant of my desires, then 2021 looked like being this particular specimen's year, so what better way to celebrate and immortalise its wonderful life than with a few well-framed portraits, I thought, as I conducted an impromptu photoshoot.

Back at home these photos would also be used to confirm I was indeed correct with my identification, as the following Sunday I posted my rock-cress and honewort pics on Twitter during #wildflowerhour to see what the great and good thought. Taking a steady stream of 'likes' to be affirmation that I was indeed correct with my honewort ID, this ground to a halt when reading a reply from an eminent young botanist, who declared my photo to be that of rock samphire instead! Mortified by this botanical faux pas, I mumbled out an online apology that as a plantsman I was still far from the finished article. But was I wrong ... ? The only way this could be sorted out would be to send my photos to the ultimate arbiter: Clive. Currently writing the *New and Updated Flora of the Gorge*, Clive's response was as quick as it was gratifying. '100% honewort,' he said ... bloody Twitter!

Having enjoyed more than a modicum of success with my Avon Gorge trip, my focus then turned south of my doorstep with preparations for an outing to catch some of the early flowering specialists across the Mendips. Starting from a pretty low baseline in terms of knowing where to track down these plants I was once more able to find help from two incredibly valuable sources: the SRPG (Somerset Rare Plants Group) website and one of their key personnel.

Representing quite possibly the best investment I would make all year, my membership of the group was already paying dividends as it not only granted access to its members' expertise but also permission to use their Rare Plant Register, a database that provided invaluable location data for many of the county's most interesting and least common plants. Simply by trawling through the database's online lists I could not only come up with a hit list of plants I wanted to see, but also work out a route by which I could catch as many as possible with a minimal amount of driving.

My other source was a stalwart member of the SRPG called Helena Crouch. In her capacity as the vice-county recorder for North Somerset (VC6), Helena's knowledge of the plants on her patch was truly exceptional. But Helena was not just a massive advocate for wild plants, she was also passionate about sharing this information, to the obvious benefit of new and keen members like myself.

It had been a couple of weeks since my trip to the Gorge, during which I'd been steadily and methodically accruing new species close to home as and when they came into flower. But with the crucial month of May almost upon us, and my grand total of just 142 different species spotted, I still felt I had a mountain to climb. If I were to try and get anywhere near the revered four-figure mark I needed to up my pace, and quickly. So with Zachary now back at school, Christina out working, Bramble already walked and a list of locations and target species at the ready, I set off for a day that hopefully would be as enjoyable as it was productive.

Having been initially cowed by crucifers as being difficult and confusing, a run of unusual and early flowering members in this family, such as shepherd's cress, hutchinsia and Bristol rock-cress, had me feeling far more confident in my abilities. This was just as well, as my first target of the day would also be my first *Draba* of the year. I just hoped it wouldn't be as dull to look at as the name of its scientific genus suggested.

Wall whitlowgrass is one of those plants that very few people without specialist knowledge would even be able to identify, let alone look for in the first place. It is a plant with a peculiar and disjunct population that sees it cropping up as a native plant of wafer-thin soils atop limestone rock, scree or walls at a few select locations along the Pennines and in the West Country. Left to my own devices I wouldn't have found this plant in a month of Sundays, but with Helena's invaluable help I'd been able to narrow my search down to a drystone wall near Chilcompton, on the edge of the Mendips.

Despite being only 20 minutes from where I lived, Chilcompton was somewhere I'd never even visited before, making me belatedly realise that it wasn't just my botanical knowledge that was drastically improving; the project was also helping to provide a greater appreciation of where I lived. Botany is wonderful like that; it allows you to explore little corners close to home that you would normally have driven straight past without a second thought. By also getting to know where I lived just that little bit better, I was also learning to love where I lived that little bit more too.

The whitlowgrass was supposedly to be found perched on an old limestone wall marking the boundary between an arable field and the verge of a single-track road. Upon spotting what I suspected might be the section of wall holding the plant, I pulled up into an old, disused gateway a short distance further along. I'd had the luxury of being able to choose my Mendip day carefully, and while strolling back towards the wall in splendid sunshine, my attention was briefly and gloriously diverted away from plants by a yellowhammer singing from a telephone cable running both above my head and along the road. This was then followed by the season's first whitethroat, with its distinctive, scratchy song revealing its presence atop a large bramble patch.

Jokingly admonishing myself for becoming distracted by birds, I got my plant-head back on as I began scanning the wall in earnest. Locating the whitlowgrass proved to be a piece of cake, with the first few plants spotted growing halfway up the wall, before the largest and best colony of around seven or eight plants was then discovered growing on top of the wall itself. Based on the somewhat decrepit condition of the field boundary, the colony had obviously been left untouched for decades, and the plants were embedded in a thin layer of soil that had certainly, in part, been formed from the decomposed remains of their predecessors. The species itself could perhaps be best described as your classic bog-standard crucifer, but upon closer inspection of the diminutive four-petalled white flowers and flattened seed-pods I decided this to be a touch unfair.

Despite being a winter annual, populations of whitlowgrass can supposedly persist for quite some time when conditions are suitable, and certainly the biggest threat to this colony seemed to come from the encroaching tendrils of a nearby bramble bush. With this location known about by probably fewer than a handful of botanists, it was in many ways representative of hundreds or thousands of similar rare plant sites that would have previously existed, before then simply disappearing due to neglect. Determining that this should not be the case here too, I decided to carry out my own bit of impromptu conservation by quickly hopping over the wall and then spending the next 20 minutes cutting back the bramble with my penknife. Many might not care if this dull and uninspiring little wall plant disappeared from this unremarkable section of wall in the middle of nowhere, but I certainly did!

My only other thought while hacking away at the invasive bramble was how this tiny colony had ever been discovered in the first place. While trying to imagine who might have initially found it, my mind conjured up the image of a Victorian gentleman with a handlebar moustache, sporting a cane and in his Sunday best. Strolling back from church, he could conceivably have spied the plant growing in exactly the same location it was currently. 'Hang on a moment, I think I might have found *Draba muralis*!' he may well have mused out loud in a perfect received pronunciation, before then informing his local naturalists' society of this important find. If the plant's discovery was even remotely like this then I was the lucky recipient of the gentleman's diligence. Good on him, I thought.

With a brief stop on the way to catch up with another crucifer called alpine pennycress, which liked hanging out on the Mendips' heavy-metal slag heaps, my main target species for the day required a visit to one of the numerous limestone woodlands running along the hills' south-western slopes. Purple gromwell was undoubtedly the plant I'd previously earmarked as 'Mendips' most wanted' when carrying out my initial research. Although not entirely confined to Somerset, as small populations can still be seen in North Wales, South Wales and Devon, my adopted county had to represent the plant's British headquarters. The *Atlas Flora of Somerset* mentions the plant as being common in the

Axbridge-Cheddar area, but unlike the two crucifers I'd already seen that morning, which had been easily tracked down with grid references, for the gromwell I only had a location to go on. This wouldn't ordinarily be a problem were it not for the fact that Cheddar Wood covered a large section of the Mendips' western flank, making it potentially a needle-in-a-haystack job.

The only other niggling matter bothering me as I parked up close to the reserve's southern boundary was that large parts of the wood seemed out of bounds. While managed by the Somerset Wildlife Trust, in a peculiar arrangement it was only privately leased to the Trust, with the reserve's online webpage declaring in capitals and with some words even underlined that 'THIS PARTICULAR RESERVE IS NOT, AND NEVER HAS BEEN, AN OPEN ACCESS PUBLIC SITE. PLEASE DO NOT TRESPASS.' Despite the Trust (and the owners) having apparently previously overlooked the fact that some members of the public regularly accessed the site, the huge presence of ash dieback, in their opinion, had also made the reserve now hazardous to enter. This meant that unless I wanted to grasp the trespassing nettle once again, the only points from which I was permitted to search for the gromwell would be from public footpaths either running along the western perimeter or towards the eastern end of the wood.

By now the sun's heat was really starting to kick in, meaning the sooner I was under the trees' cooling influence the better, so I wasted little time in setting off along the steep footpath that delineated the reserve's western flank. In a day packed with a number of locations to visit, speed was of the essence, and the quicker I found the gromwell, the more time I'd have to track down some of the Mendips' other delights. Back at home I'd pored over the contours of the wood and realised the walk would entail a rise in elevation of close to 200m between where I'd parked at the bottom and the top, most north-westerly corner of the reserve. This swift ascent saw me quickly putting on a sweat as I stopped for my first breather upon reaching the wood's south-western corner.

As I took a moment for both a quick swig of water and to adjust the position of my rucksack, a splash of yellow suddenly caught my eye while glancing down at the base of a wooded hedgebank. Happy to use this plant as a proper excuse to stop for a while longer, I bent down for a closer look and within a heartbeat of spotting the blooms' imperfectly distinct petals knew I'd found a good plant. Goldilocks buttercup is fairly common in deciduous woodlands and hedgebanks on lime-rich soils, but could be one of those plants that you might conceivably go for years without seeing if you don't look in the right places and at the correct time of year. The buttercup's flowers are uniquely malformed, with some petals seemingly fine, while others are reminiscent of the crumpled wings of a freshly emerged adult butterfly or dragonfly. The fact that the buttercup even bothers to produce these crumpled flowers in the first place is even more puzzling when you realise that the plant is apomictic.

A form of asexual reproduction without fertilisation, apomixis is the domain of a small group of plants where the seed only appears to develop from unfertilised (and unpollinated) egg cells. Despite the fact that insects are regularly observed picking up the pollen before then passing it on to another flower, this genetic material remains apparently unused, with the result that all descendants – bar genetic mutations – are clones of the mother plant. This introspective attitude towards propagation has resulted in possibly up to a hundred subtly different micro-species across the country, with many of these growing in very restricted areas due to the plant's limited powers of dispersal.

For the purpose of my thousand-plant challenge I had decided early on to shy away from further splitting any of the complex species, such as the brambles, dandelions, hawkweeds and eyebrights, so this would most definitely be the one and only goldilocks buttercup that I'd be adding to my list.

Leaving the sexually sterile buttercup behind, I carried further on up the slope, and while relishing the workout my heart and lungs were receiving I was becoming increasingly frustrated that the woodland to my immediate right was simply off-limits. From the confines of the public footpath I could clearly see a wash of cobalt blue across the woodland floor created by sheets of bluebells, but I was after a much rarer plant with a subtly different hue. And if the gromwell preferred the woodland's interior to its edge, this would potentially make a tricky search even more problematic. The only other notable species added during my ascent was that of a grass. Wood melick is a specialist grass of woodland edges and rides, and which only deigns to make an appearance in our finest forests. The grass's open and drooping panicles contain just a few spikelets, with each enclosing just a single floret. Taking a moment to admire this delicately-formed species, I decided that in my humble opinion this had to be the loveliest of all Britain's grasses, with surely the only possible competition coming from the lime-loving quaking grass.

At the top of the path, and with the wood slowly petering out into meadow to my right, I was then confronted with a dilemma. Running along the top of the wood was an access track for a limestone quarry immediately to the north, and having had not even the slightest sniff of purple gromwell up to that point, I strongly felt my best chance would be in amongst the trees. This reduced my options to either ignoring the signs and trespassing, or to walking all the way along the quarry access track, before then descending along the reserve's eastern footpath, which at least would take me through a portion of the woodland rather than along its perimeter. Being instantly recognisable to many in the wildlife watching and conservation community means that potentially I have more to lose if caught ignoring signage on nature reserves, and I had no desire to be made an example of – but equally I needed to see the plant.

I decided to compromise, in that I'd walk all the way round to the other footpath, but if no gromwell was forthcoming then desperate times would call for drastic measures. Close to the access track a little further along, and where the woodland gave way to grassland, I quickly spotted a series of magenta-coloured spikes. Suspecting that access to the non-wooded areas of the reserve *would* be permitted, I carefully climbed over the barbed-wire-topped fence. While soon realising that the purple didn't belong to the gromwell, due to the habitat being not quite right, I was nevertheless thrilled, upon closer inspection to realise I'd stumbled on my first orchids of the year.

The early-purple is our earliest flowering common orchid, and can be seen from April onwards across large parts of Britain. Unlike most of our other orchids, it is not terribly fussy about the habitat to which it keeps, and providing the soil is not too acidic it can be encountered anywhere from woodlands to hedgebanks and roadsides to – in this case – grassland. Crouching down for a close-up, the first sense to kick in was that of smell as I inhaled a wonderful scent not dissimilar to lily-of-the-valley, tinged perhaps with an essence of blackcurrant. The flowers in many of the thirty or so spikes present were only just beginning to open, which perhaps was a good job, as after the flowers become pollinated they quickly change their smell to that of stinky tomcat.

Superficially at least, the densely packed spike of the early-purple is similar to a whole suite of other pinky-purple orchids, and is perhaps most easily separated by a combination of an early flowering date and its shiny green basal leaves, which are always heavily blotched. Arguably its most defining characteristic was a feature that I'd personally never seen before, due to it being tucked away below ground – with the plant's scientific name of *Orchis mascula* perhaps providing a testosterone-heavy clue. Here the plant's scientific name emanates from the Greek for 'testicle' and refers to the shape of the orchid's twin underground tubers. But as much as I'd always wanted to see these lewdly-shaped tubers, I'd never be willing to sacrifice a plant just for my own curiosity.

By now the sun was beating down, allowing me to soak up the warmth like a basking lizard while I lined up a view from my lofty position of an early-purple orchid in the foreground with that of Glastonbury Tor away to the south-east. Alongside the early-purple orchid, a few more flowering plants were then added to the list before I hopped back over the fence with my batteries recharged by both the sun and the orchid. Eventually as the dusty track wound northwards I was able to locate the entrance to the eastern footpath, which took me for the first time since my arrival into the shady and distinctly cooler wood.

Now on the only public thoroughfare through the wood, I could immediately see an array of stumps, where the Trust had removed disease-ridden ash trees in an attempt to make the footpath safer. One of the most serious issues with any ash tree affected by dieback is that branches can break away from the trunk at any time, which makes many ash-heavy woods a health and safety nightmare for those responsible for public access. Faced with the astronomic costs to make their woods 'safe', many owners and managers have had to make the difficult decision to close them to visitors rather than risk a possible incident, leaving even more sites off-limits to plant hunters like me. To make matters even worse, it has been estimated that 80 per cent of all our 125 million ash trees will eventually succumb to this virulent fungus, suggesting that our woodlands could be facing their single biggest change since the original clearance of the wildwoods in the first place.

All the more reason to find the gromwell, I thought, as I carefully scanned the undergrowth with each footstep forwards and downwards. Part of the problem was that my search entailed looking for a violet-blue flower in amongst a host of violet-blue bluebells, as my vigour, which had been renewed by the orchid, began to be replaced by anxiety. A few minor paths could be seen leading off both to my right and further into the wood, and with no sign forbidding my access I thought it wouldn't do any harm to pop in for a quick look. As I left the footpath behind with a small sense of trepidation at either being caught off-piste or risking a branch to the head, the woodland seemed almost instantly to be of a higher calibre. Surrounding this footpath offshoot were swathes of bluebells and white wood anemones, and with the steadily unfurling leaves above creating a dappled effect below, I had never seen a spring woodland in finer fettle.

It was the leaves I spotted first...

Just by my feet I noticed some long, dark green lanceolate leaves, which were shiny and downy in equal measure, and by simply tracking these leaves back to the stem, a few startling purple flowers could clearly be seen sprouting from their axils. What passed from my mouth at the moment of discovery could only be described as an involuntary gargle of delight. Out of the 149 plants I'd so far successfully found and identified, number 150 was comfortably the best of the lot. It was stunning; it was rare; I'd never seen it before; it was close to home and it had made me work for it – this was a plant that ticked all the boxes.

I furtively looked around to make sure no one was watching me watching the plant, and satisfied I had the plant to myself, I did something I'm not sure I've ever done with a wild plant before – I lovingly fondled it. Apparently when the funnel-shaped flowers of purple gromwell first emerge they are of a much redder colour, before then turning the more-familiar intense blue, and in the few plants I subsequently found there was not a hint of red to be seen. This mattered not, as I admired the flowers in what could only be described as pristine state. Once this terrific plant had smiled for the camera, I quickly back-tracked the short distance to the main path, before then floating back down to the car. I still had 850 plants to find, with plenty more lows and spills inevitably lurking around the corner, but for now I was going to enjoy the highs and thrills, because I'd just found a botanical bobby-dazzler.

CHAPTER 6

Chilterns-bound

As April turned into May I began picking up the pace. My technique developed into using the daily dog walks to ensure nothing common or widespread was missed as it came into flower, while targeting reserves close to home that would offer something different in between. Sand Point, just north of Weston-super-Mare was one such example. Here, a family outing to the beach was seamlessly combined with a walk up to the point, which eventually revealed the tiny but perfectly formed suffocated clover in the highly cropped turf, along with a few sprigs of wild clary garnished with a freshly laid dog turd, both of which were not just 'lifers' for me, but also valuable additions to the Dilger database.

Covid also seemed finally to be on the back foot too, and with a government announcement on 17th May rescinding advice to stay local, this meant that the Chilterns trip I'd long been hoping to undertake could now go ahead.

Comprising a range of hills stretching diagonally across England from the River Thames in southern Oxfordshire to Hitchin in Hertfordshire, the 840km^2 that comprise the Chilterns were finally recognised as an Area of Outstanding Natural Beauty (AONB) in 1965. While this rural landscape has been shaped by people for centuries, as nearly two-thirds is farmed, in between the fields of wheat, oat and oilseed rape are large blocks of predominantly beech woodland which also contain smaller pockets of unimproved grassland. And it is these two valuable habitats that offer far higher levels of botanical diversity than the cropped monocultures that frequently encircle them.

What crucially enables the Chilterns to punch above its botanical weight is that the underlying bedrock across large swathes of the AONB consists of chalk. In the absence of any agricultural improvement, such as the addition of fertiliser or manure, the chalk here only tends to produce thin calcareous soils, which favour an array of special plant communities. The Chilterns are particularly famed amongst ardent plant-twitchers for an impressive suite of rare and threatened orchids. While already acquainted with the work of a couple of young botanical guns, who'd both written extensively about their attempts to see every single British orchid in one season, I had no desire to follow in their footsteps. It would, however, have been remiss not to log at least a few of the finest and rarest members of this charismatic group.

When asked what my favourite month is, my smart-Alec response usually tends to be 'the second half of May and the first half of June'. This transition between spring and summer is when the bird-breeding season is at its height, badger and fox cubs appear out in the open for the first time and hedgerows are plastered with hawthorn blossom. It is also probably the period when most species of plant choose to bloom and hence was a month when I fully expected – and hoped – to score heavily.

With Zachary now back at school and Christina in full-on gardening mode, some work in the East Midlands allowed me to tack two full days of wall-to-wall botany onto the front of this paid gig while barely needing to make a detour. More importantly, Christina had given the trip extension her blessing. Having always travelled for work, packing my bags is as much second nature to me as brushing my teeth, but the enforced period at home, as we all played our part in helping to stem the pandemic's spread, certainly gave me a strangely unfamiliar feeling as I loaded the car before setting off for the Home Counties.

Having compiled a hit list of plants I most wanted to see, I'd planned a dot-to-dot itinerary of some of the Chilterns' best sites, in what according to the maps appeared the most fuel-efficient manner. An overnight stay in the nearby town of High Wycombe after the first day's plant-spotting would then mean I could hit the ground running for most of the second day before my presence was required elsewhere. The first planned location was Hartslock, a reserve towards the south-westerly end of the Chilterns. Managed by the Bucks, Berks and Oxon Wildlife Trust (BBOWT), the reserve consists of 10 hectares of sloping unimproved chalk grassland that not only offers great plants but also affords magnificent views of the River Thames in the valley below.

It was a site I had visited before with an old friend of mine, Tim Sykes, who I'd first met when we worked for the North Wales Wildlife Trust up in Bangor. Being a far more accomplished botanist than me, and originally hailing from Oxfordshire, Tim had been keen to show me a few of the Chilterns' rarest orchids, and the magic of that trip had stayed with me for the 25 years since.

Due to its rural location, Hartslock is not one of the easiest reserves to reach. The Trust advise walking the final mile and a half from the nearby town of Goring along the Thames Path below. However, with such a packed itinerary, I decided to ignore their advice by parking as close as possible to the reserve's eastern end, which by my own calculation would halve the distance. Walking down, I was pleased to confirm the weather forecast appeared to be right, with the temperature in its low teens as blue sky and scudding clouds battled for supremacy. Certainly my chosen route in was definitely the road less travelled, meaning I made a couple of mistakes while skirting around oilseed rape fields and a mountain bike track along the way. But this gentle amble downhill also presented the opportunity to warm up my botanical brain by picking out a whole raft of species I'd already caught up with earlier in the year. This botanical equivalent of singing the scales would hopefully mean I'd be in full voice by the time I hit the reserve proper.

The surprisingly unassuming display board that marks the entrance to the reserve belies the fact that you are about to enter one of the most celebrated chalk grassland reserves in Britain. Had Hartslock been famed for its birds rather than its plants there would doubtless have been both a reserve centre and a cafe selling refreshments. But – if you'll pardon the pun – the bunting is generally only hung out for birders, as they outnumber botanists many times over. This lack of interest in our flora can of course be both a blessing and a curse. On the one hand you can often visit a premier plant reserve at the best time of year and frequently experience the secret pleasure of being the only person there. But on the flip side, because so few folk care about these sites, the scant publicity they tend to garner means that relatively few voices are heard if they become, for example, suddenly threatened with development.

Orchids, however, are the one family of plants that seem to be the exception to the rule. Their beauty, in many cases rarity, and restricted number of species give them a kudos that can attract even the most plant-blind of naturalists. I personally know a number of birders who can barely identify the difference between a dandelion and a daisy, but have seen virtually all the British orchids – which takes some doing! So it was perhaps of little surprise upon entering the reserve that for once I wasn't the only person who'd come to admire its botanical treasures. The first patch of 50 or 60 orchids wasn't that difficult to spot, as finding them entailed nothing more than seeing where four or five other botanists were either lying prone, while taking photographs, or chatting amiably alongside. If the other folk hadn't been inadvertently signposting where to look, then the fenced-off rope and hastily written sign appealing for visitors to stay on the paths made their discovery even more straightforward, and if I'm honest a touch anti-climactic.

The reserve is most famous for its population of monkey orchids, one of only three sites in the whole country, and crucially by far the most accessible. In addition to a number of more common orchid species, such as bee, pyramidal and common spotted, a lady orchid of unknown provenance has made an occasional appearance here as well. Having seen both monkey and lady before, I noted that this colony appeared to look like neither. These orchids were substantially larger than I knew monkeys to be, and while equivalent in stature to the far stouter lady orchids, their collective colour scheme was definitely unladylike. The individual flowers of lady orchids tend to have a crimson bonnet, which contrasts with their pale spotted lip, but the bright cerise pink colour of these plants was more reminiscent of another rare orchid, the military orchid, which I knew not to be here ... so which species were they?

Casting a furtive glance around at the other botanists present, it appeared they seemed perfectly content with what they were looking at, which made me suddenly feel a touch self-conscious. Needing help, I surreptitiously retrieved my BFF field guide, before thumbing through to the orchid pages.

'They're hybrids between lady and monkey,' said a voice, which belonged to a lady possibly 15 years my senior and who had obviously noted the vexed look etched across my face. Mumbling my thanks back, it suddenly all fit into place. Orchids can be notoriously promiscuous when it comes to hybridising between closely related species. So common is this with marsh orchids in the genus *Dactylorhiza*, for example, that species boundaries can end up becoming quite the moveable feast, with the result that those tasked with botanical nomenclature are often making name changes to reflect this taxonomically fluid state of affairs.

It also explained why the specimens in front of me looked so healthy and substantial. Often when two closely related species hybridise, a process called hybrid vigour takes place, which sees the resultant offspring taking on the most desirable traits of both parents. In this case, these 'lonkeys' or 'm'ladies' had somehow taken on the robust physical stature of the ladies but retained the gorgeous colouring of the monkeys. The lady – whose name was Hilary – and her husband Steve, having noticed I was receptive to more of their pearls of wisdom, then went on to inform me that this was the only place in Britain where the hybrid between these two species has ever been recorded since its first discovery in 2006.

Even more remarkably, the original single lady orchid, which had provided 50 per cent of the hybrids' genes is now rarely seen flowering at Hartslock, as the bold and brashy hybrid has become the dominant form. This hybridisation could potentially represent the process of speciation in action, whereby the population of 'pure' monkey orchids still present might ultimately be wiped out, due to being constantly crossed or re-crossed with the hybrids. DNA work carried out on Hartslock's monkeys has shown the population to have a very narrow genetic base anyway, due to the population of orchids having crashed to just seven spikes before the orchids' precise ecological requirements had become better understood by the Trust. So perhaps this injection of genes from a closely related species might even help by making them more resistant to droughts or cold snaps in the future? Only time would tell.

Slightly bending one of my own self-imposed rules, I decided that as the hybrid appeared to have occurred naturally and was so markedly different to its two parents, this meant I could allow it onto my list as a species in its own right. Chatting more to Hilary and Steve, I told them about my mission to see a thousand shades of green in a calendar year, to which they suggested I should visit the nearby Homefield Wood, for its military and fly orchids. Thanking them for the 'gen' I explained that this reserve was already firmly on the agenda for the following day, with Warburg Reserve and Dinton Pastures Country Park to be visited later that afternoon – but not until I'd seen the still-pure monkey orchids.

Sure enough, and a touch further down the sloping bank, the more diminutive monkey orchids were duly discovered. Quickly identifiable by the spindly and distinctly curved 'arms' and 'legs' of each flower and the slightly chaotic feel to the flowers' position on each spike, the presence of such an immense rarity would normally have filled me with immense joy. But with the show having already been stolen by the hybrid, and other plants to track down before the day was out, this was not a plant over which I could afford to dawdle.

As I waved farewell to Hilary and Steve, who were busily admiring a pair of mating dingy skipper butterflies, they enquired if I needed any directions for Homefield's orchids the following day, as the 'flies' were apparently quite tricky to find. I replied that as the thrill of the chase was half the fun, I'd prefer to try and track them down myself. But upon my suggestion that we should keep in touch via email for later in the season, Hilary proposed that she'd send me an annotated map with more precise details on Homefield's flies for use just in case of an emergency.

My next destination was another BBOWT reserve called Warburg, some eight miles north-east as the crow flies, but distinctly further when following Oxfordshire's rural, winding B roads. Ten times the size of Hartslock, Warburg boasts a far higher diversity of habitats and, in addition to flower-rich chalk grasslands in the valley bottoms, also contains a large proportion of ancient beech woods that rise up the valley sides. This was another of the suite of reserves I'd visited with my pal Tim back in the 1990s and with most of my other Chilterns targets lined up for the following day – apart from a slim chance of green houndstongue – I primarily wanted to use the visit as a trip down memory lane. This was the first place I'd ever seen fly orchids in my late twenties, and I was keen to see what I remembered of my initial visit all those years ago.

The reserve itself must surely be one of the remotest in the whole of southern England and can only be accessed by a tiny rural lane leading north from the delightfully named Bix Bottom. Being a more commonly visited reserve than Hartslock, undoubtedly due to its year-round appeal, the reserve is not only the proud custodian of a car park, but also a small but perfectly formed visitor centre, which due to Covid looked like it had been closed for months. Rather than obsessively tracking down species here, I wanted to just enjoy the site for its flowery loveliness, and upon walking directly north up to the reserve's grasslands I soon became immersed in cowslips, violets, buttercups and an array of the more common orchids. The woodland edges were productive hunting grounds too, with the delightful woodland umbellifer of sanicle added to my list, alongside other ancient woodland indicators like sweet woodruff and yellow archangel.

Having gleaned very little information on where best to look for the houndstongue, I just let my feet take me in whichever direction looked the most appealing. So I wasn't unduly perturbed, when via a long, circuitous route that took me up hill and down dale, I eventually found myself back at the car park without even so much as a sniff of my primary target. I reminded myself that my mission didn't involve trying to see every single British plant; sometimes it was better to take off the blinkers and instead sacrifice a few species for a more botanically holistic point of view. While Warburg had been an opportunity to soak up the Chilterns' vibe, my last site of the day would be quite the opposite, as a visit to Dinton Pastures would be nothing more and nothing less than a bare, naked twitch.

Summer snowflake has to be amongst the most poorly named of any British plants. In the first instance, it will have finished flowering well before high summer arrives and its white, nodding bell-shaped flowers are certainly not like any snowflake I've ever seen. Mostly confined to wet meadows and copses along the River Thames and River Loddon, despite its naturalisation at a number of other sites, the plant is also a member of the lily family, perhaps making its other name of Loddon lily far more suitable.

Country parks are usually best avoided for those with a strong natural-history bent. Originally designed as public green spaces for those dwelling in nearby urban conurbations to enjoy recreation and exercise, their wildlife diversity has often been sacrificed at the altar of access. They are able to cope with large numbers of visitors and often have good car parks, toilet facilities and cafes, but these are not features that rare or easily disturbed wildlife tends to favour.

One of around 400 such sites dotted around England, Dinton Pastures Country Park was originally conceived to service the towns of Reading, Wokingham and Bracknell. Here, the boating lake, play park and Dragonfly Cafe undoubtedly soak up many of the visitors, but I was keen to drop into one of the park's lesser attractions. With the River Loddon also passing through the park, the attraction for me was the seasonally inundated meadows, which ran along the watercourse at certain points.

Parking by the curiously titled Museum of Berkshire Aviation, I decided, on this occasion, to give the exhibit a miss, as I headed due south and through a little wooded copse of willow and alder in the direction of the Loddon's west bank. By now the weather had clouded over somewhat since the morning's orchid-spotting, and I was glad I'd exchanged walking boots for wellington boots as the path got progressively wetter.

In no time I was ankle-deep in water, as Loddon's floodplain did its job, but this didn't bother me in the slightest as huge sprays of Loddon lilies soon came into view. The best displays were underneath the trees, in a watery woodland swamp, with the spectacle much lovelier than I'd perhaps imagined beforehand. On each plant, the lily's green strap-like leaves were topped off by delicate, drooping umbels of between three and six nodding flowers. These blooms had the appearance of gently hanging bells, while each flower's six snow-white tepals were indelibly stamped on their outer side with a delicate lime-green smudge. Much like bluebells, where the lowermost flowers are the ones to both open and set seed first, each plant had one or two flowers already going over, making me glad I hadn't left it any later. In a few days' time the spectacle would have lost much of its appeal.

Widely grown as an ornamental, the cultivated variety is able to thrive in surprisingly dry locations, which has made the plant's spread all the easier following its escape from innumerable gardens across southern England. With so many populations now settled in our countryside, it can be tricky to differentiate between our true, native species and the escaped cultivar. The easiest way to be sure that you're appreciating the real McCoy, of course, is to catch up with the plant in one of its original locations – such as along the banks of the Thames or Loddon – as I had done. But if you are still unsure of the plant's provenance, then a simple taxonomic trick to separate the wheat from the chaff involves nothing more complicated than using your thumb and forefinger. Here, by simply feeling up the flower stem, you can identify any Loddon lily that is the genuine article as it will possess minutely toothed edges. Confirming this to be the case here too, and delighted with this equivalent of a botanical slam-dunk I splashed back to the car, happy in the knowledge that I had underestimated the ecological worth of Dinton Pastures Country Park. Perhaps more importantly, the delightfully underrated Loddon lily was thriving under the rangers' careful stewardship.

Suitably refreshed following a night at High Wycombe's Premier Inn®, and after a snatched breakfast, I headed for the first destination in what promised to be another day of high-octane botany. Homefield Wood is another fine reserve owned and managed by the redoubtable BBOWT, with the undoubted star being its population of military orchids.

Once considered extinct in Britain, around 40 military orchids were rediscovered in 1947 by the famed orchid enthusiast J. E. Lousley, who was picnicking at a site recently felled of trees as part of the war effort. He then kept the site's location secret to prevent marauding plant hunters from visiting the site with a trowel, until it was stumbled upon by the botanists Richard Fitter and Frances Rose (with the former being the first F in my BFF field guide), in 1956. Thrilled at the find, they then sent Lousley a postcard to advise him that 'the soldiers are at home in their fields.' Today the orchid is found (like the monkey orchid) at a mere three locations, with two of these being in the Chilterns, while the third is at a reserve in Breckland, where it was only discovered in 1955.

The military orchid is one of a suite of 'edge of range' species in the United Kingdom. It has an extensive range across Europe, and while confined to chalk downland within the United Kingdom, the seemingly more suitable conditions on the continent enable it to thrive in the marshlands of southern France, for example.

At just six hectares, Homefield Wood is even smaller than Hartslock and comprises a small area of grassland and adjacent forest that is set within a much larger plantation now managed by Forestry England. After parking up at the forest entrance, catching up with the orchid consisted of nothing more arduous than a short walk up a path, while serenaded by a stock dove, before then passing through a narrow belt of trees. In the meadow beyond, the spikes could clearly and immediately be seen sticking out the grass like sore thumbs. Many of the orchids seemed to be marked with a short cane, presumably for counting purposes, while a number were also covered with chicken wire. This latter measure was, I presumed, to prevent the orchids from being nibbled off in their prime by muntjac deer, which, having escaped from stately parks such as Woburn in the 1920s, have run amok across the Chilterns. With so little other suitable habitat present, I guessed this must have been the very clearing that Lousley had chosen for his lunch 74 years ago. And with the orchid's flowering period generally quite short, it could almost have been to the day too.

Following in the footsteps of these plant hunters was quite thrilling, and as I ducked down to the orchids' level a quick glance around revealed, much to my surprise, that I was in fact the only person in the meadow. This strange feeling of solitude whilst out botanising once again aroused mixed feelings. On the one hand I had the orchids all to myself – which was lovely I had to admit – but on the other hand I couldn't believe how many people were missing out.

Turning my attention back to the orchids, a quick assessment of the thirty-odd spikes present revealed the uppermost flowers to be still firmly shut, suggesting they were still a week off their absolute best. But as the flowers lower down had already opened beautifully, with each bloom revealing the soldier's pale helmet and far stouter 'limbs' than were present on the previous day's monkeys, this minor quibble could in no way detract from what was a fine display.

Bagging the military orchid had been a piece of cake, but despite 'the fly' being much the commoner of the two orchids, I intuitively felt it would be far trickier to track down. Unlike the military, for example, which prefers to make an ostentatious statement with loud colours out in the open, the fly orchid prefers hanging out along woodland edges, where the dappled shade can more easily mask its drab colouring.

Working out my plan of attack, there were at least three paths leading into the adjoining woodland at the far end of the clearing and after a quick 'eeny, meeny, miny, moe', I selected the widest and most central route. With no fly orchids immediately obvious at the junction of the two habitats, I proceeded up the path at funereal pace, while hoping to spot one peeping out from the path-side herbage. As the wood was composed primarily of beech, which always casts a deep shade, I figured the best place to locate the flies would be close to the woodland entrance, where the ambient light levels were higher. But by about 50m in, and with no orchid in sight, the light levels looked too low for the orchid and so I retraced my steps back to the meadow before then selecting the path immediately to the left.

The same drop-off in light upon entering the wood occurred here too as I quickly reached what felt once again to be sub-optimal habitat. Realising that valuable time was being wasted, my pragmatic side urged me to stop messing about and just open the email on my phone that Hilary had sent overnight. However, the idealist in me wanted to 'discover' the flies myself, much as Lousley himself would have done with the military orchids all those years ago.

After a quick internal debate, during which I told myself that if I wanted to find every single one of the thousand species myself then I would fall well short, I decided to look at Hilary's map. But due to a lack of signal on the phone this option quickly disappeared too. With the power of hindsight, had I downloaded her directions at the hotel beforehand I could then have used them as a back-up. Another option would be to find some signal, which would have involved a somewhat chastening trudge back to the car, before then driving around the area in a bid to find reception. Deciding I'd rather fail than resort to this tactic, I retreated to the meadow to try the path just to the right of my first choice.

Nothing, nada, zip ...

Back down in the meadow again I was staring down the barrel of failure. I could neither find the fly orchids nor open Hilary's email; it was pathetic and hilarious in equal measure. Casting my eyes around, perhaps more in desperation than in hope, I could see one last possibility in the form of a barely distinguishable path also running up into the forest, but virtually alongside the road I'd originally turned off to reach the car park.

Knowing I was in the last-chance saloon, I walked over before then finding at least half a dozen flies within just a few seconds of entering the wood. Finally spotting them felt almost anti-climactic, but this lasted only as long as it took for me to take a closer look. If orchids were to be given personalities then the fly would perhaps be labelled as 'introverted'. But still waters often run deep, and admiring the flowers' velvety chocolate bodies and steely-blue waists, I could instantly appreciate why the author Richard Mabey had commented that fly orchids 'resemble wingless bluebottles impaled on stalks'.

Timing-wise, I hadn't quite nailed it with the flies either, but plenty of the dozen or so plants I eventually found along the path had at least one or two flowers open. Despite resembling flies, the flowers are in fact designed to attract a couple of species of digger wasp. This dark art of deception is achieved by the orchid's ability to produce a pheromone that is so chemically similar to that emitted by the female wasps that any males in the neighbourhood are simply unable to resist the plant's sirenic charms. Tricked into climbing onboard, the male digger wasps then inadvertently pick up pollen while trying to mate with the flower. Belatedly realising they're flogging a dead horse, the male wasps then fly off only to be duped again by another Machiavellian flower – thus completing the loop. Unsurprisingly, given the chain of events that must happen for fertilisation to occur, only around 20 per cent of flowers are thought to successfully set seed. And when a lack of suitable woodland is added into the mix it is perhaps unsurprising that the fly orchid is one of our fastest disappearing flowers.

All too aware that time waited for no botanist, I quickly headed back to the car for the second of three locations I had pencilled in before being required elsewhere. Coralroot was one of the plants I'd targeted beforehand on my Chilterns' hit list. Closely related to cuckooflower, but with slightly darker coloured flowers, this was an enticing plant thanks to the presence of extraordinary little purple bulbils in the axils of the leaves. It was also a plant I'd never seen before – unpicked that is.

Just before my trip, Christina had taken Zachary and Bramble out for a walk in Smallcombe Woods near Bath, where she had happened to find the very same plant growing under some beech trees. Innocently picking it for me to inspect, she described my face as a combination of horror and amazement when she later asked me what I thought it might be. Having recognised the plant instantly, I had not even been aware of its presence in the West Country until subsequently discovering in the *Atlas Flora of Somerset* that a small population of coralroot had survived in the woods following their suspected escape from the grounds of a large house nearby. Wanting to see the plant in its original, natural and best habitat, I had decided instead that it could wait until my Chilterns visit, with all the online research pointing to a nearby reserve called Gomm's Wood as the best place to catch this exceptional crucifer.

My satnav informed me that the onward journey was a short half-hour back east and just beyond High Wycombe. Owned and managed by Wycombe District Council in conjunction with Natural England, the wood's online blurb unequivocally stated that the 18-hectare site comprised a diverse mix of chalk grassland, semi-natural woodland and scrub – with a population of coralroot to boot.

Unlike all the other Chilterns reserves I'd visited so far, Gomm's had a much more suburban feel, due to being hemmed in by a road to the east and on all remaining sides by the High Wycombe suburb of Micklefield. The only other connection to greenery was via a small 'neck' at the reserve's northern end, which acted as a corridor to a much larger reserve called King's Wood.

The reserve was accessed by an adjacent cemetery, where it was delightful to find many of the gravestones wreathed by flowering crosswort. Having spent all morning in quite splendid isolation at Homefield Wood, I discovered there would be no such peace and solitude at Gomm's. Here the sounds of birdsong and silence were largely drowned out by a combination of traffic noise from the nearby road and the hustle and bustle emanating from a large secondary school a few hundred metres further down the valley. By way of compensation, the sun was by now well and truly out, and along the reserve's large, open rides numerous brimstone and orange-tip butterflies could be seen taking advantage of the step-up in temperature. The mature woodland on either side contained plenty of ancient woodland indicators too, such as sanicle, sweet woodruff and yellow archangel, and I was able to grab a new grass for my list in the form of wood millet. However, upon walking up to the junction with King's Wood, where the reserve leaflet intimated the coralroot grew in the dappled light, I discovered, after a brief and unsuccessful search, that I might soon be re-entering *Groundhog Day* territory.

Eventually declaring this particular location to be a coralroot-free zone, I headed back south to see if I'd simply missed them elsewhere. The intervening patches of grassland were much more than just thoroughfares to the other wooded sections, and possessed plenty of spots where the underlying bedrock could clearly be seen piercing the thin calcareous soils. Here, the usual suspects of common rock-rose, salad burnet and yellow rattle all did their best to lift my increasingly gloomy spirits as I felt all the morning's fly orchid anxieties flooding back with a vengeance. But there was to be no redemption from the reserve's more southerly beech woods either, as I drew a blank across the entire site.

When preparing for the Chilterns trip, I'd originally planned to visit one more site after Gomm's – that being a reserve called Knocking Hoe National Nature Reserve. This was a well-known location for burnt-tip orchids, but if the coralroots didn't turn up quickly, then this location would simply have to be dropped.

Now hot, bothered and carrying a rucksack that appeared heavier with each passing step, I decided to make one last attempt where I'd originally searched, along the boundary between the two reserves. On the way back I could hear what sounded like a large group of children chatting excitedly from a small glade I'd noted earlier, containing a circle of seats around a fire-pit created for school groups. From their voices I estimated the children to be around four or five years old, but not in the mood for polite conversation, I decided the best course of action would be to put my head down while passing, in an attempt not to be seen.

Most appeared to be looking at the teacher and not in the direction of an irate botanist trying not to be noticed as he strode past, but one sharp-eyed child did catch sight of me and shouted in a loud voice, 'Look!...There's an old man!' With incredulity I suddenly realised he was looking at me. True – due to tramping up and down the woodland my face had become a touch flushed, and I am follically challenged, which can age men prematurely, but on the flip side I've always considered myself younger than my 56 years and certainly in good shape. Upon hearing the child's proclamation the whole class then turned round to see where he was pointing, as the teacher too looked up with an apologetic smile. As everyone stared at me like I was some kind of weird curio, I suddenly became so incandescent with rage that I wanted to march up to the child before scolding him for what I felt to be insolence.

But upon realising that my plant-hunting, on occasion, ironically brought out the grumpy old man in me – and with my anger quickly subsiding – I chose instead to quietly slink off into the undergrowth.

Calling off my fruitless search I retired back to the welcome shade of my car to take stock. I might have been an old dog, I thought, but at least I was an old dog with a bone. The one point of difference between Homefield and Gomm's was that the latter's suburban location at least meant I had a phone signal, and upon googling 'Chilterns and coralroot' once more, was astonished to see that the very first hit related to the plant's presence at a 'Gomm's and Bubbles Wood' a few miles away. Had I visited the wrong Gomm's Wood? The leaflet at the Gomm's I'd just left clearly contained a picture of coralroot, but after the best part of two hours' searching, I could only conclude it had either been lost to the site, or had become so rare that finding it in the time left would be nothing short of impossible. Gomm's 2, however, was also a site managed by the Woodland Trust and the reserve's online management plan clearly stated that coralroot was present.

With insufficient time now left to visit both the burnt-tip site and Gomm's 2, this left me with a dilemma as to whether to stick or twist. While the burnt-tip is undoubtedly one of our loveliest orchids, it was also a plant I'd seen a couple of decades before, whereas the coralroot was a lifer. Coming to the conclusion that it had also become personal with the coralroot, I decided to chance my arm at a site I'd never even visited before, let alone researched. Gomm's 2 was nestled in the Hughenden Valley, and my map seemed to indicate that it was situated in a far more rural location than the Gomm's I'd just left, meaning it would also be far less likely to contain impudent name-calling children. Access also appeared easiest from the wood's north-western tip, which was adjacent to an independent day school for girls, called Pipers Corner.

Relieved to finally see High Wycombe disappearing in the rear-view mirror, I was soon back in the rural Chilterns countryside, and parked down a small lane that appeared to mark the dividing line between the school and the reserve. Spotting a path with a Woodland Trust display board, I noted that if I were to carry straight down into the wood then I'd hit what appeared to be the main path.

Instantly the wood felt different. Being mostly composed of large mature beech trees, a dense shade cast by the canopy above resulted in the sparse ground cover so typical of many of the Chilterns' best beech woods. With no one else to be seen, and during that lull in bird activity that often occurs before the dusk chorus picks up, I was delighted to find that I would have heard a pin drop. I was also pleased to find the wood a few degrees cooler than Gomm's 1, and after descending to what appeared to be the main thoroughfare, quickly found a small patch of coralroot growing no further than a cricket pitch's length away.

It's a peculiar feeling when you eventually see something you've been straining every sinew to find. Rubbing my eyes to make sure I was indeed looking at the object of my desires, I could clearly see the four rose-pink petals of the single flower atop a slender, straight stem that contained a nuggety little purple bulbil in the axil of each leaf. Expecting to have been euphoric at the find, the overriding emotion instead was that of relief. This gorgeous little specialist of the Chilterns and Sussex (and a few woodlands in and around Bath) also represented the perfect full-stop to my Chilterns trip. Species no. 236 had been duly spotted, meaning I was just shy of a quarter of the way to my target.

Despite the gloom on the woodland floor and my lack of a camera flash, I decided that I would at least try and take a few snaps, which entailed me having to lie prone, half across the path and half in the undergrowth. Just me and the plant, exactly how I liked it…

Suddenly there was a scream. Startled by such a loud and piercing cry that appeared so close, I accidentally dropped my camera while recoiling from whatever might be about to attack me. Quickly realising that I was not about to be bludgeoned, I then craned my neck around to see at least half a dozen girls in uniform, presumably from the nearby school, standing stock still while looking on, aghast. Seeing their upset, I mumbled that I was photographing the plants, to which one of the girls stammered that they thought I was dead.

Fancy being called 'old' and 'dead' in the space of just a couple of hours! This plant spotting was certainly taking its toll.

CHAPTER 7

Plate-spinning on the Lizard

When busily planning my big botanical year at the end of 2020, the very first task on my extensive to-do list was to arrange a trip down to the Lizard Peninsula in Cornwall. The Lizard, as it is frequently known, is Cornwall's 'second toe', and represents the most southerly point of mainland Britain. Amongst the many visitors to west Cornwall however, it often plays second fiddle to the 'first toe' of Land's End, which is not just England's most westerly point but perhaps best known as the starting point for those wishing to walk or cycle the 874 miles to John O'Groats. But to naturalists with even the remotest interest in flora, Land's End is considered relatively dull when compared to the botanical Mecca of the Lizard, visible just across the water of Mount's Bay.

The remarkable diversity and richness of rare plants on the Lizard is principally down to a couple of key factors. Although the geology across the peninsula is incredibly complex, the southern half is dominated by one type of rock: serpentine. This dark green and red rock, which looks like shiny reptilian skin when weathered, was originally formed over 600 million years ago, before then being squeezed up from deep within the earth's crust during a collision between two continental plates some 300 million years later.

Almost entirely confined to the Lizard, this highly unusual rock gives rise to soil characterised by a number of fascinating properties. Whilst very low in calcium, it has high levels of magnesium. This means many 'calcicoles', or calcium-demanding species, are absent, while many lime-loving species are more than able to cope. But conversely many 'calcifuge', or calcium-hating species typical of more acid conditions, are able to thrive here too. These serpentine-derived soils also tend to be of a shallow and skeletal nature, giving the plants a more precarious hold than in, say, deep, rich and loamy soils.

Due to the Lizard's extreme southerly position, the climate plays an equally important part in the composition of its flora. Being surrounded by sea warms the peninsula to such an extent that the Lizard frequently boasts the highest mean daily temperatures in January and February of any location on mainland Britain. So for the plants anchored here, this invariably results in frosts being rare, while the growing season is long. Rainfall is relatively low too, with much falling in the winter, while droughts in spring and summer are relatively common. These elements all combine to give the Lizard an almost Mediterranean feel, where plants with a predominantly southerly distribution have made the Lizard their most northerly outpost.

As 20 Nationally Rare 'Red Data Book' plants (established by the International Union for the Conservation of Nature, or IUCN, to safeguard and protect our rarest plants) and many more Nationally Scarce species can be found across the Lizard, it ranks as the single richest area in Britain, alongside Ben Lawers and its adjacent mountains, for rarities. Amongst a number of species found nowhere else in the United Kingdom, the Lizard's famous trio of clovers were right at the top of my 'must see' list. Although upright clover has occasionally been seen at Stanner Rocks (where I caught up with the Radnor lily), the other two members – long-headed and twin-headed clover – are entirely confined to the Lizard. In fact the botanical gem of Caerthillian Cove, on the Lizard's south-western tip, supposedly holds a grand total of 14 species in the genus *Trifolium*, making this small valley the best location for clover-spotting in the whole of Britain.

A botanical pilgrimage down to the Lizard for the majority of the rarities is considered best in late May to early June, and with Zachary's half-term break commencing on the last day of May, it was a no-brainer. After all, who doesn't like a beach holiday? As the nation had been deprived of vacations due to Covid for the best part of 15 months, accommodation at many holiday hotspots had quickly sold out. So I was thrilled to be able to reserve a dog-friendly static caravan at a small remote campsite, close to the seaside village of Coverack in the Lizard's south-eastern corner.

Packing the car with everything from buckets to botanical field guides, I felt excitement and trepidation in equal measure. Family holidays are always a balancing act, where 'compromise' has to be the watch-word, given all the participants' competing interests. Zachary, for example, would want to spend most of his holiday on the beach, and while Christina would be keen to explore some of the Lizard's many coastal paths and cafes, I was anxious to spend every available moment with my nose a couple of centimetres from the turf. These three activities were not of course mutually exclusive, as many of the best botanical sites are frequently close to the beach and right alongside coastal paths. But in order to both find and identify the clovers, I would need intense focus and minimal distractions, which might prove easier said than done.

With me putting these mildly disconcerting thoughts to the back of my mind, we set off, happy in the knowledge that the long-term forecast looked decidedly improved, considering the unseasonably cold weather we'd been subjected to during the first half of May. Even from Bristol, which is considered the gateway to the west, it can be a long slog down to Cornwall, and as our caravan wouldn't be ready to receive us until later in the day, I'd planned for our first holiday stop to be that of Marazion Beach. Just east of Penzance, and looking over the famous St Michael's Mount in the Bay, the long, sandy beach is backed by sand dunes, which I also knew to be a good spot for coastal plants.

Upon parking, we loaded up with assorted paraphernalia before opting to build our base for the day at a section of beach close to where the Red River emptied into the bay and adjacent to a portion of the extensive dune system running away to the west. Walking through the sandy and partly vegetated hummocks to our chosen spot, I could already see a tantalising array of plants awaiting my attention, but these would simply have to wait as family interests came first. If I'd chosen to abandon their needs at this early stage, in favour of plant-spotting, then a lynching would possibly be on the cards.

There was so much here for Zachary to do that he was initially spoilt for choice and so divided his time between paddling, building castles and madly dashing around the beach with Bramble in tow. Mercifully the sun was shining too, and with the temperature indicating T-shirt weather, I suddenly felt that tingle that you only tend to get when you are on holiday. To add to our delight, while the beach was certainly well-attended it was by no means too busy, giving everyone more than enough room to socially distance.

Next up on the day's packed itinerary was a beach-comb for Christina and Zachary, while I manned the base. And after the assorted shells, cuttlefish bones and sea potatoes collected from an especially productive strandline had been admired, my time to search for plants was duly granted. Being wedged in between the coastal road on the landward side and the beachgoers' trampling feet on the seaward side, the dune system at Marazion could perhaps best be described as small, but perfectly formed. This can often be problematic for such a dynamic, mobile habitat that often resists all attempts to be hemmed in, but in this case the dunes' minimal coverage did not appear to have impacted their floral diversity.

The first flower I'd clocked earlier, and which now required a quick consultation from my BFF field guide, was a gorgeous aster-type flower that had painted certain patches of the dunes a delicate pinkish-white. Looking just like supercharged daisies, the blooms, I wasn't surprised to discover, belonged to the same genus as Mexican fleabane, which I'd previously seen plastering many of the walls in the Chew Valley.

Originating from North America, seaside daisy has become yet another colourful addition to Britain's multicultural flora, courtesy of the international trade in plants. First recorded as growing wild on Bournemouth's sea cliffs in 1942, it has subsequently colonised a number of locations across the south coast, in addition to a few locations elsewhere. Perhaps the most puritanical of botanists would have considered the daisy, being an immigrant, a 'lesser plant'. But as we'd brought it here in the first place, it could hardly be blamed for such opportunistic tendencies, and the plant's bright, perky flowers had certainly brightened up my dune stroll. Having also decided early on that naturalised plants were perfectly acceptable, I was only too delighted to add another alien to my list.

Further along the dunes, the native sand dune specialist that is sea holly was somewhat more familiar. Not quite yet in flower, the plants were sprinkled in a long, thin line, just behind where most of the beachgoers had staked their claim with towels, and would have undoubtedly prickled a bum or two had they been just a few paces closer to the sea. As I wandered happily from flower to flower, it became obvious that Marazion's dune flora was every bit as diverse a mix of native and alien species as you might encounter growing in our urban environments. The only difference here was the composition of species – as kidney vetch, common storksbill and sea beet all combined to form a complex mosaic alongside the sprawling Japanese rose and the long-since-introduced charlock.

I was just at the point of turning around when a flower suddenly popped into view that was so stunning it almost took my breath away. Immediately recognisable as a bindweed, due to the plant's prostrate nature, its fabulous candy-pink and white flowers instantly brought to mind a strawberry Cornetto®. With pink seemingly the predominant colour, each of the flower's five petals was then bisected by a white stripe radiating from the centre outwards. Unsure if I'd ever seen sea bindweed before, whether or not the plant was another lifer for me did little to detract from my moment of elation. I then recalled that this hadn't even been the first bindweed species we'd recorded that day, as before leaving home Zachary and I had harvested a small supply of hedge bindweed leaves from our garden to ensure our tortoise would be well fed for the duration of our absence.

The last plant of interest to be spotted before my presence was required elsewhere, comprised nothing more than a rosette of bluish-green strap-like and twisting leaves. My initial thought was that they appeared vaguely reminiscent, albeit on a larger scale, of the wild daffodils I'd previously seen with my in-laws in the woods around Bath. Quickly scrolling through my own mental botanical database for other possibilities, it dawned on me that despite never having seen this plant before, I knew its identity. Prior research is always a key component on any trip I ever carry out, and in many ways learning more about what I might see is almost as exciting as finding it out in the field. In this case I'd come across a website listing the flora of Marazion, amongst which was a fascinating plant called sea daffodil.

Belonging to the rarest category of all our flora – that of a 'new native', sea daffodil had apparently been first reported from Marazion in 2006, where it has subsequently flowered and fruited every year since. The small colony here, which consists of no more than five clumps, is now considered a native, as the species is believed to have colonised quite naturally, courtesy of seeds washed over from known colonies of this plant in north-west Brittany. More commonly encountered on the beaches of the Algarve in the Mediterranean, sea daffodil also belongs to a select suite of plants that can withstand both the soaring temperatures of summer and the desiccating effects of a life in sand. Being only found at two other sites, in Devon and Dorset, made this surely one of our rarest plants. While thrilled with such a terrific find, the only shame was that because its flowers are only produced in high sumer, this meant that I wouldn't be able to see its striking white daffodil-esque flowers.

With the campsite now beckoning, we all decamped back to the car park for a celebratory strawberry ice cream while I regaled wife, son and dog with a blow-by-blow account of my exciting finds. But one more botanical surprise was to come in the form of a host of bright fuchsia-pink spikes dotted along the roadside as we headed onto the Lizard. Aware that the only native species of gladiolus in Britain is confined to the New Forest, I was thrilled to later find out that a second species has since been added to the British floral list. Eastern gladiolus, or 'Whistling Jacks', is another plant that originates from the Mediterranean, but having hopped over the garden wall, it too has seemingly found the Lizard's amenable climate to be the perfect home from home.

As we woke up to a cold, clear morning, my suggestion of spending our first full day at Kynance Cove was met with a chorus of approval. As a destination, this picturesque spot ticked all our respective boxes: there is a lovely walk down to the cove, a world-class beach at the bottom, a cafe with hot drinks and cake close by and a plethora of immensely rare plants studded all around.

And judging by the busy nature of the National Trust car park perched on top of the cliffs and above the cove, it seemed like we had not been alone in our thinking. Hopefully, with the exception of any time spent on the beach, we would be able to keep as far from the madding crowd as possible, I mused, as we followed the well-trodden path that runs parallel to the cliffs before dropping down to the cove. The riot of colour produced by the clifftop flowers on our descent was impossible to ignore. Thrift was easily the most numerous of all the flowering plants, and adorned virtually every promontory, with the shockingly bright bloody cranesbill and kidney vetch both contributing to a display that would have competed with anything the Chelsea Flower Show could offer up for 'best in show'. While having seen all these plants on many occasions before, it was particularly exciting to also see the bright red form of kidney vetch studded in amongst the egg-yolk yellow of the far more common form. This variety, called *Anthyllis vulneraria* var. *coccinea*, is only commonly encountered on the coastal cliffs here in Cornwall and in Pembrokeshire.

The pebbly beach at Kynance is one of the best places to appreciate at close quarters the famed serpentine rock. But as we approached the final flight of steps, we could also see a slightly more alarming sign stating that dogs were not permitted on the beach between Easter and October. This could have been a huge spanner in the works, were it not for Christina's brilliant suggestion that I should take Zachary to go and look for some of the cove's flowers first, while she sat with Bramble at the cafe. I could then switch places with her as she took Zachary onto the beach for a paddle. With the tide due to fall as the morning progressed, this order of botany first and beach second would also enable us to enjoy more of the beach's fabled golden sand as it became revealed later on.

Sitting with Bramble, a cappuccino and a book, Christina declared herself all set as Zachary and his grateful father climbed up the set of steps hewn into the rock and immediately to the west of the beach. Representing a continuation of the coastal path that leads initially west, before then turning northwards towards Mullion Cove, the terrain on either side of the path here is considered one of the richest hunting grounds. As he scrambled up ahead of me, I was delighted to see Zachary was well up for the challenge too, and equally thrilled that it was he, in fact, who found the best, and first plant, as he pointed out the dainty little white stars of spring sandwort. Emanating from a small cushion of vegetation, which was itself anchored to the exposed rock by nothing more than a shallow slither of soil, the plant's precarious toehold here enabled me to properly appreciate for the first time what an unforgiving and specialised habitat the Lizard was.

Considered nationally scarce and listed by the IUCN as 'Near Threatened', spring sandwort is also a member of a small collection of plants called metallophytes, which have developed an exceptionally high tolerance to soils contaminated with heavy metals. Apparently the sandwort is able to excrete any accumulated toxins through specialised pores in its leaves, which are then washed back into the soils. Any metals not expelled can alternatively be simply stored in the plant's cell walls instead. Spring sandwort's tolerance to lead in particular sees it growing on the spoil heaps of old lead mines across Britain, such as in the Mendips, closer to home. However, in Kynance's serpentine-derived soils, the heavy metals that must be dealt with are chromium and nickel, and judging by the abundance of the flowers here, the sandwort was patently thriving in an environment that would see many other plants struggling to survive.

As spring squill, sheepsbit and sea mayweed were all quickly identified on this most productive of slopes, we quickly found a most enjoyable rhythm, whereby Zachary searching ahead of me would be able to claim the glory of finding each new plant, whereupon I would then proceed to identify it. It was by rolling out this technique that we discovered our first mega of the trip, as the next plant the young pretender called me over to was a lime-green prostrate plant with woody stems and hairy dark green leaves occupying a small patch of otherwise-bare soil. Giving Zachary a big congratulatory bear-hug, I told him with great excitement that although fringed rupturewort wasn't the most memorable plant he'd ever seen, or the easiest to remember, it was comfortably the rarest, due to the Lizard being the only place it grew in Britain.

Being such a remote location, the Lizard's botanical importance was not realised until much later than that of the Avon Gorge, for example, with many of the endemic plants here not discovered until the early Victorian era. However, fringed rupturewort represented one of the earliest discoveries, when it was originally found in 1667 by the very same John Ray, who then proceeded to describe the Bristol rock-cress some 19 years later.

Taking an eye lens to this stellar rarity, I was able to show Zachary the particularly hairy 'eyelashes' along the fringes of the leaves, which represent the plant's key identification feature. Some of its tiny yellowish-green flowers were also open, but could perhaps best be described as underwhelming. Understandably buoyed by this top find, this then spurred Zachary on to bring my attention to yet another good plant, which he described as looking similar to eggs-and-bacon, or bird's-foot trefoil, a plant he knew well from numerous other trips out.

While not quite the level of rarity of the rupturewort, hairy greenweed was nevertheless another top-drawer find. Largely confined to the coasts of Cornwall and Pembrokeshire, the greenweed's flowers certainly look superficially similar to those of the far commoner trefoil. But the key to distinguishing the two species is to look for the greenweed's dark green, oval leaves, which could never be described as tripartite, or three-lobed. Perhaps a more closely related plant is another unusual species also present on the Lizard called petty whin, which in turn can be distinguished from the greenweed by the presence of spines.

As father and son inched up the slope, with new plants being added thick and fast, I couldn't believe that for Zachary the pull of the flowers, for now at least, appeared greater than that of the beach below. The last plant we added before tracing our steps back to Christina was also my first lifer of the morning. Spotted catsear is one of a number of very similar dandelion-like plants, which can be easily distinguished from the rest of its genre by its heavily blotched leaves. Restricted to a handful of inland lime-rich sites and a few westerly coastal sea cliffs above either limestone or serpentine rock, it represented the perfect full-stop to a fine hour's work, which had boosted my list by a princely 17 species … or make that 18 with the tree mallow we spotted growing just by the steps back up to the cafe!

Much of the second day was spent further along the Lizard's western coast at the beach at Gunwalloe Church Cove, as paddling, football and sandcastles held sway. Always keen to capitalise on any opportunity, I'd still managed to identify rock sea-spurrey, which I'd spied growing on a drystone wall near the car park, while sea rocket and sea radish were encountered on the beach. Gunwalloe was also the site where an incredibly rare sand-and-shingle specialist called sea knotgrass was rediscovered in 1966, after having been declared extinct on the British mainland four years earlier. Doubtless some buried seed had been unearthed by natural coastal processes allowing it to gain a tenuous foothold once more, but on my brief searches just above the strandline – in between games of catch – the plant's hold was to prove a little too tenuous for me, as I drew a blank.

After a somewhat more traditional day on the beach at Gunwalloe, the third day of our holiday had been inked in for some more hardcore botany. Keen to get a few clovers under my belt, I decreed that Caerthillian Cove, or the Clover Capital of Britain, would be the sole focus of the day's activities. The walk to one of Britain's most important botanical sites is nothing more than a stroll down public footpaths from the starting point of the village green in

Lizard itself, and once again the weather looked kindly as we set off for the coast. Caerthillian had been the location I'd been both most excited and also most concerned about visiting, primarily due to the fact that the clovers were not terribly easy to find and could also be devilishly difficult to identify. It was also the site I felt least prepared for, having only visited the cove briefly once before while walking from Kynance to Lizard Point back in 2008. All I knew from my all-too-brief research was that the best clover locations were on south-facing flanks of the valley where the soils were at their skinniest.

Scanning the village's rural lanes as we walked, we were soon able to add the first new plant to my list, as it busily sprouted from a crack at the base of a garden wall. Closely related to our smaller, native species, greater quaking grass, by contrast, has far larger pendant seed-heads, which give the plant a decidedly ornamental look. As the slightest of gusts made the nodding spikelets gently and delightfully tremble in the breeze, I was also able to appreciate why it had been brought over here in the first place. Following that well-trodden path from an ancestral home in the Mediterranean to Britain's garden centres, this grass now appears to be slowly increasing across southern England. It will be interesting, however, to see what happens if it turns from being a fun and quirky neophyte to our next Japanese knotweed.

Looking further down the lane, I could also see three middle-aged gentlemen, who like us appeared to be stopping on a regular basis to look at what I could only presume were plants. With the route a one-way ticket to Caerthillian, if this were indeed the case, then it could be the best possible news, as other botanists would not only have more of an idea as to precisely where I should be looking, but also what I might be looking at.

As we descended down to the impossibly picturesque cove, by now I had to admit I was keeping more of an eye on the potential help ahead of me than the plants at my feet. The footpath then began meandering along the southern side of the valley, as I watched all three botanists branch purposefully left before disappearing over a small rise and out of view. Seeing my agitated state, and realising that the clovers were quite possibly the biggest reason I'd booked the holiday to the Lizard in the first place, my fabulous wife came to the rescue once more by suggesting that I should follow while she would hold the fort with Bramble and Zachary. Pathetically grateful once again for her self-sacrificing and pre-emptive gesture, I muttered something about 'not being long' as I chased after them.

I'm always slightly wary of gate-crashing someone else's party, but on this occasion I needn't have worried, as looking up, one of the trio shouted 'Hello Mike!' as I approached the group. While recognising him, I couldn't initially put his face to a name until he introduced himself as Gareth Jones from Bristol University, whom I'd previously interviewed for a BBC *The One Show* film on noctule bats. Being someone who regularly appears on television does have its advantages, and in this case it was an immediate introduction to the other members of Gareth's merry band of botanists.

As I explained my year-long project, one of the other members, a chap called Ian Bennallick, revealed that he was none other than the BSBI county recorder for East Cornwall and had offered to guide Gareth and his friend towards some of the more interesting plants. On hearing that he'd be equally happy to assist me too, I gushed out my eternal gratitude as I realised I'd hit the botanical jackpot.

Ian went on to explain that where we were standing was the exact location the Lizard's most celebrated Botanist, C. A. Johns, had conducted his oft-quoted 'hat-trick' back in 1847. A schoolmaster from Helston, Johns had written in his book *A Week at the Lizard* that, 'So abundant are the *Leguminosae* at this spot that I covered my hat with *Trifolium bocconei*, *T. striatum*, *T. molinerii*, *T. scabrum*, *T. arvense*, *Lotus hispidus*, *Anthyllis vulneriana*.' He then went on to say, 'Had the rim been a little wider I might have included *Genista tinctoria* and *Lotus corniculatus*.' Elsewhere, Johns admitted that his impromptu quadrat had in fact been that of a 'straw hat with a broad rim', but it was nevertheless an impressive haul.

Bisecting the south-facing patch of dry grassland where we were standing was the smallest hint of a path, which looked like it had been created by generations of botanists while enjoying one of the richest floral spots in Britain. As if to illustrate his point, Ian then pointed out a patch of upright clover – one of the trio I'd been so anxious to see. Dropping to my knees for my first view of this plant, I could see it was flourishing in two discrete patches. The flowering looked to be just about spot-on as the plant's pinky-purple flowers formed neat globular heads. But it was the clover's tripartite leaflets that represented its most distinct feature. Being far narrower than those encountered on our commoner clovers, each mini-leaflet had small serrated teeth along its margins, which more closely resembled to my polluted mind the middle section of a cannabis leaf than one of our rarest clovers. It was a sublime plant.

While busily taking the upright clover's mugshot, Ian then called me over for a 'long-headed', my second member of the holy trinity. When I greedily asked him if *Trifolium bocconei* (twin-headed) was showing here as well, Ian explained that this particular clover was having a poor year. Apparently a party of Britain's finest botanists had only managed to find a couple of paltry specimens just a week earlier, making it highly likely we'd draw a blank too. By way of recompense, and before his group's departure along the cliff towards Kynance, Ian was kind enough to help by pointing out two more quality clovers – in the form of 'rough' and 'burrowing'. Having successfully identified western clover myself, just a short while after their departure, I had brought my count of new *Trifoliums* to five in just under 50 minutes. I was both physically and metaphorically in clover.

Suddenly realising that I should now rejoin my family, I scurried off down towards the valley's small beach to find a slightly frazzled Christina worn out with the responsibility of ensuring that Zachary didn't hurt himself during his rock-ninja routine, while Bramble didn't break a toe sprinting across the pebbly beach. As we split the tasks of managing dog and child I wisely refrained from telling her how my time spent away from my family had been the most rewarding part of the holiday yet.

The next couple of days were spent in full holidaying mode, as we played on the beach, went kayaking and passed a delightful afternoon at the picturesque fishing port of Porthleven. A quick trip to Lizard Point, so Zachary could briefly hold the position of 'Britain's most southerly person' was also slotted in. The drop-in to this slightly tatty tourist-trap also gave me an opportunity to catch up with a couple of the Lizard's most colourful plants. Here the yellows, pinks and purples of the aliens hottentot-fig and purple dewplant smothered large sections of the cliffs with their vast, succulent mats.

But having set my stall out to see a thousand species, there could be no resting on my laurels, so even when not actively out looking for plants I made sure I was putting in the groundwork back at the caravan each evening. My concern, for example, that the Lizard's heathland plants were still under-represented on my list, meant that once Zachary had been put to bed, my time was put to good use by trawling online for any information likely to help my cause.

Being someone who is utterly incapable of having a lie-in, while my wife can think of nothing better, I'd already managed to slip out early one morning for a quick return to Caerthillian to better photograph its clovers. Utilising this early-bird slot had proved to be a genius idea, as it allowed me to return in time for a family breakfast with my botanical itch already scratched. This also ensured that my incessant plant-spotting wouldn't impinge on sacrosanct family time. So upon the discovery of an online report by Natural England, detailing the precise location of another of the Lizard's endemics – that of pygmy rush – I started thinking that a similar dawn raid might prove equally profitable.

Of all the locations highlighted in the report as being key for the pygmy rush, I'd pinpointed a heathland site close to both the hamlet of Traboe and the Goonhilly Earth Station, primarily as it was only a 10-minute drive from our campsite and offered easy access on foot. As I slipped out of the caravan at 6 a.m., this appeared too early even for Bramble, who did little more than raise an ear as I grabbed my rucksack and headed out into a cold and clear breaking dawn.

The north-east corner of Goonhilly Downs is the largest remaining heathland block of the somewhat fragmented Lizard NNR that is managed by Natural England. Arriving at the site, I was able to properly appreciate the full extent of the flat and treeless terrain that is so characteristic of the peninsula's southern half. It is probably only here and in the New Forest that such a large extent of heathland still exists, and harks back to a bygone era, when the population was a fraction of what it is today and before agricultural intensification changed our landscapes forever. In the New Forest it is the remarkable survival of the commons-system that has kept the area relatively intact, whereas the Lizard's heaths were saved by the low fertility of their serpentine-derived soils.

While initially appearing little more than a featureless monoculture, an array of distinctly different heathland habitats could clearly be seen upon a closer inspection, which subtly morphed into one another according to the soil's depth and wetness. The section laid out before me was a fairly dry portion of heath dominated by a range of heathers, alongside an old rutted track. These numerous tracks, which criss-cross the land, are considered one of the heathland's most important habitats in their own right. Being marginally lower-lying than the surrounding terrain sees these tracks frequently flooded in winter, before then becoming baked dry in the summer. Just the kind of location where the diminutive and underwhelming but incredibly rare pygmy rush seems to thrive.

Walking across to the track from the adjacent road, I was instantly rewarded for my early rise with a new plant for both list and life. Dotted along the track, in small patches of bare ground that rose no higher than a couple of centimetres above the surrounding terrain, were discrete tufts of what initially looked like thrift on stilts. It was only upon closer inspection that they were suddenly revealed to be wild chives. The pale-violet flower heads were doused in dew, and the papery bract that had once completely enclosed the flower head could be seen gently withering below. I was particularly thrilled to notice the dark purple vein running down the centre of each petal. And confirmation of a successful identification was ensured by a quick sniff of the plant, which unsurprisingly did smell distinctly 'chivey'. While naturalised across a whole range of locations, away from its core population on the serpentine heaths, the only other sites where this perennial herb is considered native are a few discrete spots in north Cornwall, South Wales and northern England close to Hadrian's Wall.

Elsewhere down in the ditch, I also managed to identify a couple of distinct sedges, in the form of flea and yellow sedge, but upon locating a few tiny rushes, simply felt I didn't have the level of expertise to judge whether they might be the fantastically rare pygmy rush or just their common toady equivalent. With *Juncus* being such a tricky genus, I decided my time would be better spent taking a few half-decent photos rather than wasting valuable time agonising over their identity.

Having spent a lot of early spring working on the various violets growing along the hedgebanks close to home, I felt this to be a genus with which I'd become far more comfortable than that of, say *Juncus*. So upon spotting a violet that my plant 'jizz' (its general impression, size and shape) instantly told me was different to the familiar sweet, common dog and wood dog trio I knew well, decided it to be worthy of a quick identification session. While the flowers did appear to be more bluish in colouration than the more common species, this in itself was hardly the clincher until I noticed the blunt yellow spur – a feature only the heath dog violet is supposed to possess. Tentatively satisfied with my identification, I then proceeded to rise from my customary kneeling position, without fully taking into account that the blood had not quite returned to my lower limbs. The weight of my backpack didn't help either, as I stumbled forward like a drunk before face-planting some 1.72m further into the heathland vegetation.

Immediately realising I was unhurt, my first reaction was to glance around to make sure no one had seen my embarrassing tumble. But of course at this ungodly hour, and in this forsaken place, there would probably have been no one within a couple of miles from where I now lay prone in the vegetation. Somewhat closer, however, and only a few centimetres from my nose, was another violet. Different again to the heath dog, the flower was a creamy pale lilac, making it look like a white flower that had been given the most delicate rinse with a watercolour paintbrush. Once again the spur was different, being shorter and more greenish than I'd just seen on the other violet, and it also appeared to possess longer and narrower leaves.

Pale dog violet is a nationally scarce plant that is listed as 'Vulnerable', due to the fact it appears to be declining everywhere. Mostly confined to southern and south-westerly sites in England, with additional strongholds along the Welsh coasts of Pembrokeshire and Anglesey, it is apparently made rarer still by its predilection for swapping genes with heath dog to form what's known as a hybrid swarm. This plant, however, appeared to have none of the vigour commonplace amongst hybrids I'd seen elsewhere, such as the monkey-lady crosses in the Chilterns, nor intermediate characteristics – it seemed to be a genuine pale dog violet.

Getting out of bed early can sometimes help find a totally unexpected and rare plant. Plus with all the other 'dog' violets logged during my quest, I'd now compiled enough to pull a sled.

As I arrived back at the campsite, everyone was tucking in to a full Cornish breakfast as I regaled them with both the plants I'd found on my planned trip and also my unrehearsed trip. For some strange reason, Zachary was far more interested in the blow-by-blow account of how I'd taken the tumble rather than the finer identification points of *Viola lactea*, but that's eight-year-olds for you. With our last day sadly upon us and home calling, I'd planned one last excursion for the Chew Stoke Botanical Group before we turned for home – a visit to a National Trust property north of Land's End. Boscregan is a tenanted farm owned by the National Trust and managed specifically for rare arable weeds, and few come rarer than the farm's star attraction and my principal quarry – purple viper's bugloss.

Whether purple viper's bugloss is considered a native plant or an archaeophyte (an ancient introduction) is open to considerable debate, but Boscregan is considered the only place where the plant is known to grow in any abundance. Like many weeds it suffered a dramatic decline due to changes in agricultural practice, leading to just a handful of plants being present when the Trust acquired the property in 1995. Cornfield flowers are undoubtedly the most threatened plant group, with seven British species having already been exterminated in the ruthless push for food productivity. The Trust were obviously keen to ensure that the bugloss wouldn't be the eighth. The key of course to the bugloss's conservation was understanding the plant's needs, before then enacting appropriate management, and so apparently successful has this been that the plant is now considered impossible to count.

After pouring the contents of the caravan into the car we headed north and west to the delightful Nanquidno Valley, tucked away close to the town of St Just. Off the beaten track and away from the main touristy hubs, large parts of Cornwall still feel delightfully stuck in a bygone era, and the valley's tiny hamlet was undoubtedly one of those settlements with an 'end of the road' feeling to it. Having parked up, we followed the valley's wooded stream downwards, with the only plant of note being the gunnera, or giant rhubarb, which had escaped from nearby gardens, before proceeding to stake a claim along the bank-side vegetation in a couple of spots. This was a plant I knew very well from my pre-television life while working as a biologist in the Ecuadorian cloud-forests, where it had been given the local name 'pobre hombres' paragua' or 'poor man's umbrella'. As the leaves are capable of reaching well over a metre across when fully grown, I am indeed able to confirm they work wonderfully well if you happen to get caught in a tropical shower while looking for hummingbirds and toucans.

Taking a right-turn across the water to follow the footpath up the valley's steep side, we were quickly rewarded with a view across towards both Land's End and the Longships Lighthouse away to the south-west. Moving steadily away from the water, the trees slowly petered out, as the habitat changed to a mix of heathland and low-intensity pasture grazed by a herd of particularly impressive longhorns. These hardy beasts are increasingly being used as a conservation grazing tool to prevent grasslands and heathlands from scrubbing over. At the sight of these mighty bovids, both Zachary and Bramble appeared cowed, but seemingly regarding our sudden appearance with little more than fleeting interest, the cattle kept a respectful distance as we passed on by.

Of course, with nature conservation, balance is everything, and with a certain amount of scrub being considered essential for nesting birds, an understanding of stocking levels plays an important part. If the head of cattle placed on the land is too high, for example, then everything will be nibbled and trampled to oblivion, but with sympathetic management, a diverse mosaic can be achieved. Coming across a bank of western gorse I was suddenly thankful that the Trust understood that 'less cattle was more' when we noticed large patches of the scrub festooned in what appeared to be strawberry liquorice laces. These sinuous strands belonged to a parasitic plant – dodder – and it was obviously having a field day.

Capable of sprawling over any plant, but favouring gorse and heather, dodder is a parasitic plant with a fascinating lifestyle. Able to detect the closest suitable hosts by the chemical signature they emit, the dodder seedling enacts 'Operation Sprawl', which comprises sending out numerous red tendrils to ultimately twine in anti-clockwise fashion around its victim's stems. Once the host has been successfully snared, the dodder then proceeds to tighten these coils before structures called haustoria are produced with the precise function of penetrating the host in order to siphon off any nutriment the dodder needs. When it does find a large patch of gorse, the dodder can proliferate incredibly quickly, and although the host plant can be weakened it is rarely killed.

One downside of our arrival so early in the dodder's life cycle was that we'd be denied the opportunity of smelling its pongy flowers, which supposedly smell of rotting fish. In the eyes of an eight-year-old, there is only one type of plant more fascinating than one which commits daylight robbery, and that is one that also happens to reek to high heaven.

As we moved up towards the farm, the purple viper's bugloss was clearly visible along the field margins from some distance away, making it by a clear distance the easiest plant to locate of the entire trip. As it was still relatively early in the bugloss's flowering season, only the advanced guard had begun to bloom, so we wouldn't have the opportunity to experience the purple and yellow haze from the viper's bugloss and corn marigold that the farm is famed for in high summer. But what's not to like about a rare plant in a beautiful location? And moreover, it was another plant to add to the list from a very profitable week.

'Chow! Chow!' I immediately looked upwards, for possibly the first time all holiday, as we watched a party of six choughs fly straight past us while heading for the sea cliffs below. Plants might have been the focus all week, but birds still had the capacity to turn my head … and particularly a species as charismatic as the Cornish crow.

CHAPTER 8

In Revered Company

With June already upon me, I had little time to waste. And so less than 24 hours after arriving back home from the Lizard I was off again, this time across to the other side of England. Earlier in the year I'd successfully pitched an idea to make a programme for BBC Radio 4's *Costing the Earth* (CTE) strand about the impact of climate change on Britain's plants, and so where better to start the broadcast than the unique landscape straddling the Suffolk and Norfolk border – that of Breckland.

I had suggested to the programme's producer that Breckland would be a good location for a number of reasons. Firstly it is not only one of the driest places in England, but also possesses a cold continental-style climate which sees ground frosts potentially occurring in any month. These climatic quirks perhaps make Breckland's flora more susceptible to the vagaries wrought by climate change than at many other locations. Secondly I had secured the services of a superb contributor with an intimate knowledge of the Brecks, in the form of the botanist and author Simon Harrap. And perhaps most importantly (and selfishly) of all, I knew Breckland to have a remarkable flora that is home to at least 200 species of conservation priority, with a frankly astonishing 12 species found nowhere else in Britain. In short, the visit would give my plant list the most enormous fillip.

While it was impractical to reach East Anglia any other way than by driving, the Breckland visit would form the first leg of a huge triangular road trip where I'd spot plants around pre-arranged paid work. But one thing I'd learnt from hundreds of filming trips down the years was the need to ensure that business doesn't mix with pleasure. So, as it would be difficult to find and identify plants without seriously impinging on my ability to focus on the radio work, I was keen to arrange a botanical day beforehand.

Luckily for me, Simon had not only agreed to appear as a contributor on the radio programme but was also available to act as my guide for the field day. One of Britain's pre-eminent botanists, Simon is the author of numerous plant books, such as *Orchids of Britain and Europe*, and *Harrap's Wild Flowers*. As if producing botanical tomes is not enough, he also helps run Natural Surroundings – Norfolk's Wildflower Centre, near Holt, with his wife Anne.

Now considered a friend, he is perhaps more accurately the friend of a friend. I was first introduced to Simon by my good friend Nigel Redman, who is himself well known as a book publisher, tour leader and birder – not necessarily in that order. Perhaps his loftiest title, however, is that of guide-parent, along with his wife Cheryle, to Zachary. Moreover both Nigel and Simon now live pretty close to one another near Holt in North Norfolk, which also happened to represent the perfect base for a day trip to Breckland.

I had got to know Simon a whole lot better when Zachary, Nigel and I accompanied him for a day's botanising in the Brecks back in 2019. Having enjoyed his fine company on that occasion, I was keen to pick up from where we'd left off. Due to work and school intervening, Christina and Zachary would be unable to join me on this occasion, so I had the car to myself as I headed east on the first leg of a trip where plant-spotting and plant-talking would be front and centre.

Nigel and Cheryle are not just great friends and generous hosts but real foodies too, and it being our first catch-up since the pandemic there was much to discuss while tucking into a fine dinner accompanied by copious quantities of wine. But fortunately my thick head the following morning quickly began to recede as the three amigos met up in the

Redmans' kitchen before pointing the compass needle due south for Breckland.

Simon had arranged a packed itinerary, taking in some of Breckland's finest reserves, with the first stop being that of Cranwich Heath. Like many of Britain's most biodiverse habitats, most of the Brecks' heathland was lost in the devastating decades between the 1930s and 1970s. It has, for example, been calculated that during this period as much as 86 per cent disappeared after being either carpeted with conifer plantations or ploughed over and fertilised for the introduction of arable crops. As grazing by sheep and cattle also declined, and subsequent grazing by a large rabbit population was vastly reduced due to the arrival of myxomatosis in 1954, large areas of the surviving heath then became scrubbed over. This represented another nail in the coffin for many of Breckland's most valuable wild plants, which were reliant upon the land remaining open, well grazed and almost continually disturbed. But Cranwich Heath is also a shining example that all may not necessarily be lost, as Simon told us that the reserve had effectively been created following the removal of 75 hectares of forest, in a process known as clear-fell silviculture. With the trees gone, and a sheep-grazing regime reinstated to ensure the turf was kept short and the ground repeatedly disturbed, the results have been surprisingly rapid and enormously encouraging.

The weather looked to be kind as we entered what might have appeared to the untrained eye as little more than a poorly seeded field surrounded by conifers. But in conservation, looks can be deceiving, as this open, scarified heathland was just what the doctor (of botany) would have ordered. Simon had pre-warned us that many of Breckland's specialist plants could best be described as small and inconspicuous, but they were certainly not inconsequential, and I mentally prepared myself for the fact that I'd be passing a substantial portion of the day on my hands and knees with an eye lens.

As Simon knew the precise locations of many of the most interesting and rarest plants, tracking them down would not necessarily involve the usual drama (or excitement) I'd experienced elsewhere. With this in mind I'd asked Simon beforehand not to tell me the identity of each plant so I could try and work them out for myself. He accepted his position as botanical quizmaster with enthusiasm, gesticulating expectantly towards a lime-green splash growing along an old dirt-track. Assuming my now intensely familiar plant-inspecting position, I realised that Simon's newly appointed role perfectly suited his famous wit, which was, figuratively speaking, as dry as the sandy soil in which I was now kneeling.

Luckily my recent trip to the Lizard and some pre-Brecks research allowed me to quickly declare it as smooth rupturewort. 'Not bad for a species I'd never even seen before today,' I quietly told myself after receiving a congratulatory nod from my botanical sensei. One of only two native species of rupturewort in Britain, with the other being the 'fringed' version that Zachary and I had discovered on the Lizard, it made me suddenly wonder, given the distance apart that these two species occurred in the wild, whether I'd been the first-ever botanist to see both plants within the space of just three short days. Deciding that it was inevitable that some know-it-all had already beaten me to that particular punch, I decided instead to engage my mind more constructively by actually enjoying the plant.

Despite 'the smooth' adopting the same creeping, prostrate lifestyle and colour as its fringed relative, the apparent absence of woody stem bases suggested that the Breckland version was either an annual or biennial at best. Some of the tiny green flowers were even open in the fine weather, to reveal a set of five almost microscopic white petals. But the key distinguishing features once again were the leaves, whose smooth margins contrasted to the far-hairier fringes found on the plant's Cornish cousin. Simon's praise, in a slightly playful and mocking tone, somehow managed to convey that while I'd made an undeniably good start, sterner tests would surely come.

The next plant to be pointed out by Simon in his new, favourite game of 'test Mike's botanical competency.' was the somewhat trickier fine-leaved sandwort, which this time required a quick consultation with my BFF field guide before I felt able to confirm its identity. But the next plant had me stumped for considerably longer. Recognisable as a member of the pea family, due to the diagnostic pea-like shape of its purplish-blue flowers, its leaves consisted of paired leaflets running in parallel either side of the stem (pinnate), which suggested it might be a vetch. But nothing in my book seemed to fit, until I eventually realised I was looking at my first ever milk-vetch. Belonging to the genus *Astragalus*, purple milk-vetch is the commoner of the two species, with the other being an arctic/alpine confined to a handful of Scottish mountains. Most often encountered in short, dry and species-rich limey grasslands, it will also proliferate in both a few coastal sand dunes and in Breckland, where the underlying chalky bedrock has been covered by a layer of windblown sand. Simon then went on to tell me that the reason why it was so-called, lay in its supposed abilities to increase milk yields in livestock.

While many of the plants present were ones that I could easily identify, one of Simon's many talents came in the form of selecting those species he guessed I probably hadn't seen before. As I busily got to grips with each plant *in situ*, while photographing the most photogenic or rarest ones, Nigel's interest in such minuscule plants, by contrast, had already begun to wane. And upon hearing the lilting and evocative 'lu-lu-lu' song of a woodlark above our heads, he seamlessly switched back to his default setting of birding. Despite a recent upturn in its fortunes, woodlark is still a decidedly uncommon breeding bird across southern and eastern England, and is also in possession of such a scintillating song that its utterances would have normally stopped me dead in my tracks. But on this occasion, with my focus clearly elsewhere and my brain capable of spinning only so many plates, the birdsong had seemingly been filtered out. However, in need of a botanical break, I was grateful for the opportunity to gaze up – rather than down – and revel in another species patently benefiting from the removal of the wretched conifers.

Our next stop was into the heart of Thetford Forest and just north of the bridge over the River Little Ouse, which winds around the small Breckland village of Santon Downham. This bridge achieved fame in 1972, when appearing in the fifth series of the BBC sitcom *Dad's Army*, when Captain Mainwaring and his troops attempted to demolish it. In the sandy gravels of the Forestry Commission car park, Simon wasted no time in throwing down the botanical gauntlet of another species, which I was immediately able to put into the 'never seen before, but know it' category, in which I'd previously placed the smooth rupturewort.

Despite being reasonably widespread in dry sandy and gravelly places across Britain, bird's-foot was one of the plants I'd been really keen to catch up with. Happy to clinch another member of the pea family, I was able to admire the flower's whitish wings, yellow keel and candyfloss-pink striped banner. In fact Simon's description of these colourful blooms being reminiscent of a colourful cross-section of Blackpool rock appeared spot-on. Like so many of Breckland's other blooms, the flowers were a touch on the small side, so required a hand lens for maximum appreciation. Turning around to call Nigel over to see the plant, I hadn't realised that he was already admiring it, only from a distance of six feet away and through binoculars. Old habits were obviously hard to kick for a man who'd spent a lifetime watching feathers instead of flora!

As we wandered along the wide open track towards the railway that linked Cambridge and Norwich, exciting plants suddenly began to appear thick and fast. The word 'heath' featured heavily as familiar plants like heath groundsel and heath bedstraw were added to the list, but plenty of rarer plants were present as well. Smooth catsear represented my third member of the genus *Hypochaeris*, with both 'common' and 'spotted' having been added when on the Lizard, and it was also good to get my first cudweed on the list, in the form of – yes – small cudweed.

The next plant for Simon to test me with was minute, even by Breckland standards. But upon magnification, courtesy of my 10x hand lens, the tiny greenish and white fringed sepals of the flowers were suddenly revealed to be perfection in miniature. Needing a little help with identification from the botanical guru by my side, I eventually deduced it to be my first ever knawel – which up until 2021 would have seen me confusing this tiny plant with a long-tusked cetacean from the high Arctic. Nationally scarce and listed as 'Endangered', annual knawel appears to be yet another of those declining plants better suited to a bygone era. Undoubtedly annual knawel would have been more widely distributed when disturbed sandy soils, heathlands and herbicide-free arable fields were commonplace, but along this section of ride, it had found a welcome sanctuary.

Upon encountering such a delightful plant I felt that familiar double emotion rise in me once again: joy to have seen such an understated and underrated plant, and yet concern for its future. If annual knawel were to disappear tomorrow, then the world would of course keep turning. I'd also bet that 99.99 per cent of people wouldn't give a second thought about another inconsequential extinction. But in my opinion the world would have suddenly lost yet another tiny scrap of green and become a little more grey... finding rare and threatened plants was certainly proving to be a bittersweet experience.

As we worked the forest ride close to the railway, more shy, rare and highly threatened plants revealed themselves, almost all being plants I'd never seen before. Hoary cinquefoil and clustered clover were both scarce and declining species of dry, sandy places, while the presence of corn spurrey was almost certainly a relic from when 'brecking', or temporary cultivation of the heathlands' infertile soils was widely practised.

We took a break back at the car park to rehydrate and reflect on a productive morning's work. There would be little time, however, to recuperate, as a lady dashed over to talk to us as a matter of urgency; her son had found a snake during their cycle trip, and they needed our help. The mother and her son had been the only people we'd seen all morning enjoying the heathlands, albeit in a different manner to us, and upon telling her not to panic, we all went to see what the commotion was about. Sure enough, curled up at the base of a large log was a female adder, which had apparently been happily basking in the sun until its discovery.

Telling them they should instead be proud of their find, we revealed a little about the snake's life, before informing them that despite it being our only venomous snake, you are far likelier to be struck down by a lightning bolt than bitten by an adder. Visibly relaxing after having absorbed this pub-quiz fact, they both came for a closer look, whereupon the lady asked why we were here. Explaining we were on the hunt for Breckland's rare flowers, the lady – who apparently cycled here with her son on a regular basis – responded that she only ever came here for a ride through the trees. Though well-meaning, her response was another classic case of 'plant blindness' – that human tendency to ignore plant species as opposed to tree species, which was coupled with a blissful unawareness of their integral role in the planet's life-support system.

❁

Like Cranwich Heath, Santon Warren was another Breckland reserve rising from the carnage of clear-fell. But at this site, and following the removal of the trees, the Forestry Commission (in consultation with the wildflower charity Plantlife) then took the conservation process a step further. To counter the absence of processes that would normally have produced the all-important bare soil patches, such as cutting and grazing, the uppermost layer of soil had been scraped away from certain sections.

This process of 'turf-pairing' has also had the added benefit of stripping away much of the nitrogen deposited from agricultural and industrial activities, which has rained down like an aerial fertiliser on the British countryside for decades. Nitrogen deposition is probably the greatest threat to terrestrial global diversity that you've never heard of, coming only behind the twin terrors of inappropriate land use and climate change. It is also the least researched, but its impacts are already believed to be huge at sites like Breckland, where the floral communities have developed in a direct response to low nutrient levels. Although this stripping back of the topsoil to reveal the nutrient-low sandy substrate below has only been conducted on a micro-scale at Cranwich Heath, this draconian technique has already provided opportunities for those plants that struggle with the competition that extra nutrients bring, and has also created a more open and dynamic environment.

As I cast an eye across the largest of these scarified sections, I saw that open and barren areas collectively seemed to cover at least half the surface area, making the resultant habitat look as close to 'continental steppe' as I'd seen anywhere else in Britain. And judging by the very first plant we looked at properly, the management work looked to have been an unqualified success too. With a flourish, Simon stated that it was one of the rarest plants in Britain, as Nigel and I got up-close and personal with what appeared to be a sprawling biennial or perennial no larger in circumference than that of your average dessert plate.

The plant's stems appeared woody at the bases, before then forming a series of procumbent shoots with linear leaves along the stems, which proceeded to gently turn skywards to reveal in most cases a small terminal cluster of white-edged greenish flowers. Looking up at Simon smiling down, I stuck my neck out.' Is this the other knawel?' To this response, Simon simply added a nod to his grin. Always a rare plant, perennial knawel is almost entirely confined to Breckland, with a different subspecies present at one other locality – that of the remarkable Stanner Rocks in Radnor. Only ever recorded at a maximum of just 22 sites across Breckland, it had declined to just three sandy heaths in west Suffolk before Plantlife stepped in and started a very successful reintroduction project at a few select sites, of which our location was one. Talk about back from the brink...

With the day far warmer than we had anticipated, a liquid stop was needed in Brandon, the town that virtually straddles Suffolk's border with that of Norfolk. This break also presented us with the opportunity to drop into one of the smallest nature reserves in Britain. Situated in the middle of an industrial estate, Brandon Nature Reserve is no bigger than a five-a-side football pitch and completely surrounded on all sides by commercial properties. I'd visited this peculiar site a couple of times previously, as it is famed for its population of field wormwood, an odourless member of a largely aromatic genus of plants, which (away from a couple of naturalised sand-dune sites) is found in just three Breckland sites. Despite being much larger than many of Breckland's rarest plants, it is (in my opinion) one of the least attractive, and along with sickle medick – another East Anglian speciality known to be present at this site – represented, I'm ashamed to admit, little more than two extra ticks to my growing list.

Perhaps the most interesting aspect of our brief visit was how the two plants appeared to be faring far better outside the reserve than inside it. With both species considered fastidious specialists of nutrient-poor Breckland soils, it seemed that the skeletal soils along the roadside gutters and weedy front aprons of the adjoining commercial outlets were even more productive than the area specially set aside for them.

Suitably refreshed, Simon had promised one more Breckland site before we wrapped up for the day – that of Cranwich Camp. Almost back full circle to where we'd started the day at Cranwich Heath, the Camp was one of four sites set up in Breckland during the inter-war years to 'harden' the long-term unemployed, who at the time were considered as 'soft' or underprepared for work. This unpleasant social experiment consisted mostly of intense manual labour in return for dole money, and came to an end with the onset of World War II, when the camp was repurposed for military use.

With the last of the buildings removed by 1994, the only evidence of the site's previous use comes from aerial shots, where earthworks relating to former concrete tracks, the bases of hut platforms and a road can still be picked out. It has since proved to be exceptional for a suite of Breckland's rarest flowers that clearly found the disturbance created from marching feet and tank manoeuvrings to their liking. Conditions for its special plants were then enhanced even further, following habitat management of the site in 2010, which included scrub removal and turf-pairing.

Striding out into the middle of the reserve with Nigel and I trailing in his wake, Simon was out on a mission. On this occasion I knew what he was looking for, as he wanted to end the day with a real corker – the Breckland speciality Spanish catchfly. Waving for the slowcoaches to catch up with him, Simon hadn't just found one plant but a whole swarm of catchflies. The only shame was that we were a little too early in the season to enjoy both the flowers and the scent they produced. Supposedly its lacy, cream-coloured flowers emit one scent at night to attract moths and mosquitoes, while producing a different smell during the day to entice flies and bees.

Taller than I had anticipated, the plant consisted of a basal rosette of spatula-like dark green leaves from which arose a central stem that would eventually produce the flowers. The whole plant was stickily hairy, which has been suggested may help prevent the theft of pollen by wingless insects that are simply unable to climb the stems. Like the holly I'd seen earlier in my quest, Spanish catchfly comprises separate male and female plants, although some male plants do occasionally possess flowers with female reproductive parts.

So successful has the management work at Cranwich Camp been, that following the creation of a stripped plot, the population of catchflies mushroomed from 0 in 2011 to 2,929 in 2014. Patently this coloniser of open, bare ground in the Brecks was having a field day. Having also added 37 species to the list, with an astonishing 18 lifers, this meant I could now concentrate on producing good radio the following day without being distracted by the damn flowers at my feet.

CHAPTER 9

A Date with Destiny

There would be plenty of opportunity to rest up in the autumn, but with June such a key month for so many flowering plants, I needed to continue hitting the floral hotspots, and hitting them hard. Following a successful broadcasting day back on the Brecks for Radio 4, a subsequent half-day out botanising close to the North Norfolk coast with Nigel had produced a variety of farmland and coastal species, which I was thrilled to add to my list. Here, the pick of the crop had been prickly poppy, along an arable field just outside the village of Briston, and my second catchfly of the trip, in the form of sand catchfly on Beeston Bump near Sheringham.

Wildlife tour-leading has become a key component of my work more recently, and a long-standing commitment up in the Scottish Highlands had created an opportunity to stop off on the way at the one remaining botanical hotspot in England I'd never before visited – that of Upper Teesdale. While attempting to make my big botanical year as environmentally friendly as possible, there could be no avoiding the fact that on this occasion I'd have to drive. Bidding farewell to my gracious hosts in North Norfolk, I then traversed across to the A1(M), before heading north to County Durham and my accommodation for the night: the Langdon Beck Hotel.

Able to crowbar no more than a day and a half at one of Britain's most important reserves, and where the sum total of my knowledge stemmed from whatever literature I'd been able to devour beforehand, I would definitely need some help. And this would come from an incredibly well-qualified chap called Martin Furness, who worked as the senior reserve manager of Moor House – Upper Teesdale NNR, which also happened to be the largest site managed by Natural England.

Martin's patch was itself situated within the North Pennines Area of Outstanding Natural Beauty, where his reserve bisected the Pennine Way. Here the famed locations of Widdybank and Cronkley Fell came under his jurisdiction, and while appearing superficially like any other square mile of fell and upland pasture during the 268-mile hike along Britain's backbone, they were anything but. These two sites had been elevated to a position of botanical greatness by a unique combination of climate and geology, which in turn had resulted in a unique assemblage of plants.

Most of the surface geology of the reserve consists of carboniferous limestone, but long after it was deposited and lifted, the resultant stretching of the earth's crust caused magma to rise up from the deep before then being intruded in between the layers to form a hard, dark crystalline rock called dolerite. Known as the Great Whin Sill, this rock's resistance to weathering has resulted in the formation of outcrops at a number of places across Teesdale. Also during this process, the great heat of the volcanic material baked and altered the geology of the surrounding limestone, turning it into a distinctive, white crystalline marble known as sugar limestone.

Many of the rare plants of the area were known to occur either directly on this sugar limestone, on the unaltered limestone, along the Whin Sill itself, or in the flushed areas (or sikes) where springs emerged from lime-rich rocks. When the whole area then became severely glaciated during the last ice age, this changed the lie of the land again, as boulder clay and glacial till were deposited over the underlying rock. However, the resultant forest cover that developed over the sugar limestone and cliffs may have remained lighter, which in turn formed a refuge for some of the early post-ice-age colonisers. Finally, as the clearance of the woodland began, the ensuing treeless landscape became important for sheep farming, which ultimately provided a lifeline for those surviving plants unable to cope with either being shaded out or too much competition.

Considered one of the coldest places in England, Teesdale has long spells covered by snow, with air frosts occurring on over a third of the days of the year. It can also be incredibly wet, with an annual average of 244 days where some rain is recorded. But these challenging conditions seem to suit an astonishing variety of plants. At Teesdale it is the 'assemblage' of plants that is unique, with nowhere else hosting species from so many geographical distributions. A day's botanising here, for example, can unearth arctic/alpine plants more commonly found further north of Britain, which are rubbing shoulders with plants encountered at the Burren in south-western Ireland, the Lizard in Cornwall and representatives of southern Europe.

These assemblages also contain a number of exceptionally rare plants, of which two are confined to Teesdale – Teesdale sandwort and spring gentian. While the former is undoubtedly one of Britain's rarest plants, the latter has to be one of our most beautiful. Having never visited Upper Teesdale for filming, due to my singular lack of success in getting television executives excited about filming wild plants, I'd never had the opportunity to see the blue brilliance of spring gentians in flower. It was in essence the plant I wanted to see more than any other, and despite birds having been my prime focus for the last 40 years, I don't think a single spring has passed during this time without a desire to travel up and see this plant, only for life and work to inconveniently get in the way.

Settling in at the hotel after the long but glorious drive up through the Pennines, I began chatting with the hotel's chef in the bar, with the conversation quickly turning from her food to the reason for my visit. Most of the guests were invariably ramblers either half-way up or down the Pennine Way, and upon telling her I was here for the flowers, her immediate response was that I was too late for the gentians. Taking in my crushed expression, she went on to explain that she'd been born and bred on a farm close by and had witnessed them going over first-hand over the last couple of weeks.

I had been aware that my arrival would be towards the tail-end of their flowering season, with a number of photos on #wildflowerhour suggesting late April to late May as being when they were at their best. But surely I'd be able to find at least one flowering on 10th June? Putting these uncomfortable thoughts to the back of my mind, I tucked into a veggie lasagne before opting for an early night. I needed to be on best form for my allotted slot with Martin after breakfast.

As I opened the curtains the following morning to my one and only full day up in Teesdale, fog and grey cloud had combined to obscure what otherwise would have been a frankly magnificent view across to Cronkley Fell, reminding me that fine days could be few and far between in the North Pennines. However, at breakfast, my new friend the chef, was proving not only to be a good cook and knowledgeable botanist, but also an experienced meteorologist, as she informed me that it would improve later, with the fog lifting by lunchtime.

The meeting point with Martin was just a few miles further west, at the north-west corner of Widdybank Fell and close to the Cow Green Reservoir. The dam's construction in the late 1960s, had caused a huge outcry amongst conservationists at the time, as the reservoir's rising waters had also destroyed the lower slopes of Widdybank Fell, including some of its precious sugar limestone and associated rills (or streams). Martin was waiting for me in his branded Natural England SUV, and following a quick exchange of pleasantries, suggested that due to Covid protocols it might be easier if, instead of jumping in with him (and saving on fuel), I instead follow him down to the best spot, from where we could both walk.

As I followed him through a locked five-bar gate and along a track that roughly followed along the eastern shoreline of the reservoir, the weather looked like it was doing no more than hanging in there, with the cloud base (and a white-out) just a few metres above our slowly trundling cars. The conditions could not have been more different from the T-shirt weather encountered in the Brecks only two days previously, appearing more reminiscent of Iceland than England as we pulled up at the side of the road.

Striding across to the gravelly and stony flushes where water flowed across the exposed sugar limestone, we chatted properly for the first time as a golden plover called plaintively from somewhere higher up on the foggy fell. Martin told me in a broad accent – which to my untrained ear sounded like Geordie with a twang – that he hailed from close by, and in securing the job of warden in 2000, realised he'd found his calling. He appeared to be about my age, but looked as tough as the boots we were both wearing, presumably from having spent the last 20 years climbing either up hills or down dales.

We must have been barely a hundred metres from the road before we reached the section of the sike that Martin had been keen to show me. Here the crystalline sugar limestone could clearly be seen, and appeared just like granulated sugar when ground down by the combined scouring actions of weather and water. As we walked carefully along the edge of this flush, flowering plants suddenly started appearing from the calcareous, boggy soil around us.

The first new plant for me was Scottish asphodel. Rising no more than 6 or 7cm from where it was anchored, close to a dozen short-stalked and creamy-green flowers were just at the point of opening. These sat atop a slender green spike, which itself emanated from a basal tuft of flattened iris-like leaves. Having had the first plants pointed out to me, and with my eyes now suitably trained as to the plant's characteristics, or 'jizz', I was suddenly able to appreciate its abundance. Martin explained that the plant's main centre of distribution in Britain was in the Scottish Highlands, and Upper Teesdale – some 200 miles to the south – was not just a remnant outlying population, but the only place it could be seen in all of England.

I couldn't afford to dwell on this little gem too long, as barely a couple of metres further away the immediately recognisable bird's-eye primrose was still just about in flower. Dotted along the edge of the sike, it was much daintier than the far more familiar primulas, with a cluster of lilac-pink petalled and yellow-eyed flowers emanating from a single mealy stem. It looked just like a dinky drumstick. Unlike the Scottish asphodel that was just in the process of opening, the first primrose we'd located was just going over, but a quick search of the immediate surroundings soon enabled me to find a couple of slightly later specimens looking nothing short of pristine.

Like the asphodel, the primrose was another fan of the base-rich flushes, but unlike its Scottish counterpart was completely absent north of the border, with a range in the United Kingdom limited to the counties of Cumbria, Lancashire and Yorkshire. Despite a presence in both the Alps and the Baltic regions, quite why the primrose had not managed to penetrate either the Scandinavian or Scottish mountains was just another of those botanical conundrums that folk far better qualified than me had struggled to understand.

Looking up from the plant to see where Martin was, I could just about pick him out through the gloom and further up the sike, in that now-familiar bent stoop befitting a botanist on a mission. As I caught him up, he informed me that this precise location was considered Teesdale sandwort country, and from now on I needed to avoid standing on any of the numerous hummocks, which had been shaped into mini-islets by the running water. Martin, in what appeared to be his typical understated manner, told me to prepare to be underwhelmed, as he directed my gaze down in the direction of a single tiny white bloom. Partly encased by long bicoloured sepals at the end of a slender, dark red stem, the flower was even partly open, and upon carefully crouching down I could just about make out its protruding anthers, while its surrounding petals were flecked with raindrops.

I have always loved rare things, and as such I had to disagree with Martin's assertion that seeing it would be underwhelming – if anything it was quite the opposite. This plant was so rare that in my mind's eye it almost seemed to have a supernatural halo around it, but whether it was the rarest plant I'd seen so far on my mission was open to conjecture. As far as I could recall, the only other candidate had to be the single Radnor lily I'd found in flower at Stanner Rocks, but either way, both plants sailed into the botanical rockstar category.

In Britain, the Teesdale sandwort is only known from Widdybank Fell, where it is confined to the bases of mossy hummocks located along the very fragile stream-side margins where sugar limestone reaches the surface. Talk about exacting requirements ... It was currently only known from six localities across the fell, and recent surveys have suggested a 50 per cent decline since the late 1990s. This plant was the very definition of 'Endangered'.

'I would never have found that on my own in a month of Sundays,' I told Martin, to which he responded that he knew of a botanist who was desperate to do exactly that. Having shunned all offers to be shown the plant had meant that he was still looking. Martin elaborated that he wished instead that the gentleman would seek assistance, as having someone constantly stomping around such precious and easily damaged habitat was in no one's best interests and certainly not in the sandwort's.

Before returning to the cars, Martin was able to show me my third lifer that morning, in the form of dwarf milkwort. This member of a very familiar family is not just the smallest, but also the rarest milkwort by quite some margin. Once again this key member of Teesdale's unique assemblage has a bizarre disjunct population in Britain: being found at three sites on the closely grazed chalk turf of Kent's North Downs, inaccessible rock ledges and fissures on limestone scars across Lancashire and Yorkshire or, in this case, along flushed areas of Widdybank's eroded sugar limestone.

Unable to spend more than a couple of hours with me before his day job kicked in, Martin had promised to point me in the direction of Cronkley Fell's special plants, which I'd planned to tackle that afternoon, but before that had just enough time to show me the turfy, wet pastures right alongside Widdybank Farm. Driving up to the farm, where one of the outbuildings also doubled up as his office, I could immediately see the habitat was completely different. Whereas the walk up the sike had been at an altitude of over 500m, down in the valley we were well over 100m lower. Here the landscape was dominated by a distinct series of low-lying hillocks that had been created when glacial till (or boulder clay) was deposited by the retreating glaciers.

Martin told me that many of these pastures were pretty dull from a botanical point of view, but where springs or seepages occurred the diversity tended to sky-rocket. Splashing across the meadow, I instantly recognised hundreds of globeflower heads from quite a distance, but upon closer inspection, and dotted amongst them, were a range of plants all with something in common – that of 'marsh' in their names. While admiring marsh valerian, marsh lousewort and an early marsh orchid, something entirely different suddenly stopped me in its tracks.

I was by now approaching a level of botanical knowledge that if I wasn't instantly sure of a plant's precise identity then at least I could usually ascribe it to a family or genus, which usually made the correct attribution process somewhat more straightforward. But with this rich deep-purple and devilishly handsome species I wasn't initially sure at all. 'Alpine bartsia,' said Martin, having noticed me gazing downwards in slack-mouthed awe. I briefly berated myself for not having realised it belonged to the broomrape, bartsia and cow-wheat family, before actively pushing these negative thoughts to one side. My time would be far better spent enjoying the plant instead.

No taller than a 15cm ruler and intensely hairy all over, the erect stem held pairs of toothed bugle-like leaves at regular intervals, with a short terminal spike holding most of the rich purple flowers, where each bloom emerged from a dull-purple coloured bract. The arresting purple colour of the flowers almost lent the plant a sheen of crushed velvet, which in my opinion elevated it to the level of any British orchid in the glamour stakes. Everyone should remember where they were when they saw their first alpine bartsia – I certainly would.

When not encountered in just a few wet meadows flushed with base-rich waters around the North Pennines, the only other place where it can be tracked down is on a few mountain ledges in Perthshire and Argyll. Even here in the shadow of Natural England's office, which must surely rank as one of its best-protected, well-managed and closely monitored sites, the TLC showered on this arctic/alpine had not prevented it from joining the same inauspicious category to which the Teesdale sandwort firmly belonged – that of a rare species in a seemingly terminal decline.

Martin gave me a rundown of the bartsia's precise requirements. Firstly the grazing and trampling offered by cattle and sheep was vital for keeping the lusher vegetation down and the habitat open, but crucially it also needed to be at the right level. Apparently too much nibbling and stomping can lead to excessive compaction of the soil and subtle alterations in hydrology. For this most fastidious of arctic/alpines to have the opportunity to flower and set seed, the livestock also needed to be removed during the summer months. Quite the effort – but certainly worth every penny of my taxpayer's money.

Unable to accompany me to Cronkley Fell, Martin was, however, more than happy to point out on my map the best route up, with helpful suggestions of locations and species to look out for along the way. Despite being thrilled with my Teesdalian haul so far, one species had still been notable by its absence – spring gentian. When I asked Martin if he knew of any location where it might still be flowering, he agreed with the hotel chef's assessment that the plant's best days had undoubtedly passed, before then offering a glimmer of hope. Apparently the colder climate and harsher conditions up on Cronkley Fell's summit often meant that the gentians in the exclosures were frequently the last to flower across Teesdale – and I'd take a glimmer any day.

Driving round to Forest-in-Teesdale after a hastily eaten lunch, I was pleased to see the chef's weather predictions were proving remarkably accurate, and while not exactly blazing hot, it did at least look like one of the 121 days in each calendar year when the rain would keep away.

Taking a footbridge across the River Tees, I was welcomed back onto the NNR by a Natural England noticeboard as I followed the footpath south along the well-marked Pennine Way. On this occasion I must have passed only a couple of folk along this usually well-trodden route, while also spotting the likes of water avens, cuckooflower and marsh marigold in the path-side ditches. It was fascinating to see how far these early spring plants were behind their compatriots further south. Marsh marigold, for example, always looks its best in late March or early April in my own Chew Valley garden, in Somerset, while early June appeared to be its time to shine in cold, wet and windy Teesdale. Despite my focus being primarily on plants, old habits die hard – making it impossible to ignore the birds. And as I left the road and river behind, the sound of the fell began alternating between the lilting, downward cadence of innumerable willow warblers and the bubbling of breeding curlews – the most complimentary and delightful upland pairing anyone could possibly wish for.

Departing the main Pennine trail, I then turned uphill in a westerly direction and towards the summit of Cronkley Fell. Here I was relieved to see that the high ground ahead appeared mercifully clear of fog, and despite being windier than I'd expected, alternating glances between the sky above and the weather app on my phone seemed to indicate that this state of affairs would continue for the rest of the day. As I steadily climbed, the views became ever more panoramic, allowing me to properly appreciate the reserve for the first time. Apparently the management of Upper Teesdale is a delicate and complex business, given that none of the reserve is actually owned by Natural England. Within its boundaries are four tenanted working farms, with the rest either grouse moor or sheep walk and whose businesses must be balanced with the conservation of the reserve's flora and fauna. On the one hand this made looking for rare plants somewhat easier, as the best botanical areas were generally within easy-to-spot fenced exclosures designed to keep out the sheep. But on the other hand, as the remaining chunk of fell was so intensely grazed, this resulted in its natural history interest being minimal.

With so little of botanical note as I approached the fell summit, the birdlife did an admirable job of keeping me entertained. The plaintive two-toned whistle of the golden plover provided a near-constant soundtrack, making me wonder why this was a bird I never heard in an area of upland far more familiar to me – the Scottish uplands surrounding Cairngorm Mountain. But the highlight of this steady slog upwards was provided by a male ring ouzel, which sang its heart out from a drystone wall, with an appreciative audience of just one – me. Otherwise known as the mountain blackbird, the ring ouzel has to be my favourite upland bird, and a species that is always the focus of an annual pilgrimage to the Brecon Beacons. In South Wales the small population of ring ouzels seem to be responding to climate change by retreating ever higher up the hills with each passing year, so at some point could theoretically run out of upland. But certainly here in Teesdale, I was able to enjoy a bird that appeared perfectly in sync with its environment, as I was reminded for the second time that afternoon that once a birder, always a birder.

Hopping over the fence of this initial exclosure, I was suddenly surrounded by flowers for the first time since leaving the valley. Out of the reach of the sheep's incisors, these plants would at least be able to complete their annual life cycles without being nipped off in their prime. Here, the purple form of mountain pansy, bird's-eye primrose and Scottish asphodel all jostled for space in the skeletal turf, and while catching my breath I was also able to spend a delightful five minutes differentiating between common and hoary rockrose – the latter being the infinitely rarer of the two. In the exclosure it was also strange to see horseshoe vetch, a plant I more closely associated with England's hot southern chalk downs, thriving alongside hummocks of spring sandwort I'd only seen elsewhere thriving in the Lizard's baked soils. The bluish-purple of blue-moor grass was common here too, but one flower seemed notable by its absence: spring gentian.

While the best time to enjoy Upper Teesdale's flora can vary from year to year, mid May to early June is generally considered the window that most botanists will aim for. But my visit on 10th June, which was the earliest I'd been able to manage, certainly appeared to have been too late for the spring gentian class of 2021. Normally such a distinctive flower would have stood out clearly, but picking out the plant's vegetative parts alone would be far more difficult given my modest level of expertise. All the more reason to return at another time, I told myself as I desperately tried to put a positive spin on such a crushing disappointment.

Hopping along the various exclosures, I was able to pick out more unusual assemblages, with one containing heather, common rockrose and wood anemone in one patch, three plants that I'd traditionally associated with radically different habitats: moorland, calcareous grassland and woodland, respectively. What a wonderful and bizarre place Cronkley Fell was. Due to the lack of any shops nearby, the hotel had sorted me out with a packed lunch, and as I reached for my first sandwich buried in my rucksack right next to me, I suddenly noticed a tiny scrap of blue.

The gentian's famed propellor-like petals were firmly closed and undeniably on their last legs, while the calyx was in an even worse state and withering in front of my eyes. The plant was by now just a shadow of its former glory, but it was a shadow I'd happily take. With my eye now firmly in on the plant's 'jizz', I spotted a few more close by that were even further gone, to the extent that no blue could be seen at all, but I had undeniably found Teesdale's most famous plant. The county flower of Durham, one of my most sought-after species and in an astonishing place to boot – life was good.

Following the path further west, I eventually began dropping down off Cronkley Fell, with the plan to follow the Tees' riverside footpath all the way round the north of the fell I'd just climbed until arriving back at my starting point. As I descended, with Widdybank clearly viewable across the valley and the River Tees below, I could also see a huge exclosure away to my left, which demanded further investigation. As I arrived at the fence line, the difference in the state of the vegetation either side could not have been more stark. On the side I was standing, the vegetation had been grazed to within an inch of its life, but on the stock-proof side, I could see extensive stands of juniper and enough ferns to have reduced any pteridophyte-obsessed Victorian plant hunter to a gibbering wreck.

A selection of boulders no further than 20m beyond the fence line had created a shady crevice, providing shelter from the prevailing wind, and in no time I was able to add the delightful parsley fern, oak fern and hard fern to my ever-expanding plant list. Due to the elimination of cattle and livestock, the exclosure looked like it was already halfway along the process of turning into a juniper woodland, while on the other side of the fence line the high stocking rates had taken it too far the other way – surely it shouldn't be beyond our wit or wisdom to find some kind of a middle ground? It's a tough life for the sheep up on Cronkley Fell and equally hard, I'd imagine, for the farming communities working the land up here, who must rely on subsidies just to make ends meet. I'm not the first or last person to remark on this, but why not pay the farmers instead to manage this incredibly important land solely for wildlife? Surely that would be the ultimate win-win?

CHAPTER 10

Re-forming the Band

A week spent leading a wildlife tour up in the Scottish Highlands immediately following my Teesdale visit allowed me to maintain the momentum, although as the week was primarily focussed on birds and mammals, any plant sightings I did manage were largely opportunistic. For some bizarre reason the guests appeared keener on seeing crested tits than cloudberry and divers than dwarf cornel, but nevertheless I did manage to persuade them that a few charismatic species, such as twinflower and coralroot orchid, were worth a quick detour.

On my return home, a second day trip out with the Somerset Rare Plants Group – this time to a calcareous mire in the south of the county – took me closer to the halfway mark. In frankly appalling weather, we managed to track down a few species I'd never seen before, like pale butterwort and fen bedstraw, while also benefitting from the group's expertise on some of the more difficult taxonomic groups. In the short periods spent at home during the rich floristic period of late June, dog walks doubled up as impromptu botanical trips, with my 500th species logged being meadow cranesbill, which was spied at the foot of a hedgerow less than five minutes from our back door. Yet despite the numbers already seen and the obvious improvement in my botanical knowledge, I still felt daunted by the challenge remaining.

Early on in the planning process, I'd decided that my mission must not focus entirely on plants, but should also feature people, and in particular some of the fabulous folk I'd met during the course of my previous conservation and current television careers. While continuing to include great plants, these excursions would also offer the opportunity for a different kind of trip – one down memory lane. Reconnecting with old friends and colleagues after an absence of (in some cases) years would also be great fun.

When planning my big year, my old botanical buddy Tim Sykes was one of the first I reached out to. Tim and I became acquainted while working in Bangor for the North Wales Wildlife Trust, and he was now based down in Hampshire and employed by the Environment Agency with a remit to protect the county's fisheries and enhance its biodiversity. A few years older than me, Tim had a ready smile and twinkling eyes to go with his big frame and short-cropped blond hair. Additionally, his genial nature always won over anyone he met. We instantly got on, and became firm friends as we bonded over a shared passion for wildlife, which in my case was birds, while Tim was all about plants.

Having previously been employed at the Brecon Wildlife Trust, Tim had famously been airlifted off the cliff face at Craig Cerrig Gleisiad in the Brecon Beacons while becoming stuck during a botanical hunt for purple saxifrage, and he had brought that same spirit of (mis)adventure along with him to his new job in North Wales. Part of Tim's rationale for moving from Brecon to Bangor was to explore Gwynedd's fabulous array of plants, which encompassed everything from the arctic/alpines in Snowdonia to dune specialists on Anglesey. Sensing an opportunity to round out my skill set beyond a knowledge of just birds, I began tagging along on some of his excursions.

Tim had a mischievous sense of fun and once surreptitiously slipped a couple of boulders into my rucksack while we climbed Snowdon, with me only realising exactly why I'd struggled with the weight of my rucksack once at the summit. On one trip to the Great Orme we almost managed to recreate Tim's Brecon Beacons rescue while looking for dark-red helleborines, while another to Tim's family seat in Oxfordshire had resulted in the fabulous 'orchid-centric' Chilterns trip that I'd emulated back in May.

Above: Coltsfoot at an abandoned – but jealously guarded – brownfield site in Bristol.

Left: Halfway up a rock-face, a solitary bloom of Radnor lily immeasurably brightens up a dull, cold day in February.

Below: Purple gromwell can be difficult to spot when hidden amongst a 'sea' of bluebells.

Above: The delicate, pendant flowers of Solomon's seal, just about to open on a woodland floor in Somerset.

Left: Identified by its spindly 'arms' and 'legs', the mercurial monkey orchid is confined to three chalky grasslands in southern England.

Below: A speciality of the Chilterns, coralroot is easily identified by its bulbils.

bove: Long-headed clover is one-third of
ıe Lizard Peninsula's stellar trio of clovers.

ight: Perhaps more familiar as a garden herb,
ıe Lizard's heathlands are the stronghold for
ild chives.

elow: A trio of botanists enjoying the many
elights that stud the skeletal serpentine soils
f Caerthillian Cove, on the Lizard.

Above: Perennial knawel is the recipient of a very successful reintroduction project.

Right: Finding many of Breckland's small and inconspicuous plants requires both intense focus and the help of a terrific botanist – like Simon Harrap.

Below: A denizen of sandy places, bird's-foot resembles a cross-section of Blackpool rock.

Above: Parsley fern thriving in a 'sheep-free zone' on Upper Teesdale's Cronkley Fell.

Left: Intensely hairy, rare, velvety-purple and incredibly striking - everyone should remember their first sighting of alpine bartsia.

Below: Teesdale sandwort is one of our rarest wild flowers; despite a modest bloom, it should never be described as 'underwhelming'.

pposite, clockwise from top ft: A single mousetail flowers amongst both cows and eir pats in the New Forest. onfined to a few Caledonian ne forests and plantations, ıe-flowered wintergreen is vine, distinctive and darned re. Surely our rarest rockrose, early visit is essential to catch ıe spotted' before it sheds its tal. Seaside centaury is tough d terrific in equal measure.

lockwise from top right: eadow clary is one of our ost striking wild blooms. ooking and smelling like miniature pine seedling, ound-pine is a plant of a gone era. Catching up with orwegian mugwort requires rseverance and perspiration. pine milk-vetch is under reat from grazing, trampling d climate change.

Right, top to bottom: Confined to Scottish mountain ledges, alpine fleabane looks like a hairy, pink and supercharged daisy – and doesn't even grow in the Alps!

Rising out of the pond's murky depths, just like Excalibur, starfruit is so scarce that it has become a plant of mythical rarity.

Below, left to right: While not quite as stunning as its 'spring' or 'alpine' cousins, the bright blue flowers of marsh gentian are definitely worth getting your feet wet for.

Going for gold…father and son enjoying species number 1,000 – otherwise known as goldilocks aster.

I'd also been honoured to take up the position of Tim's best man when he married his fiancée Josie, but after he moved to the Environment Agency our communications became a touch more sporadic as he rose through the ranks, while I ploughed my own furrow in television. Re-establishing contact to see if Tim would be willing to re-form the band for another gig, this time close to his own base in the New Forest, I was delighted when he declared it a splendid idea and a date was set in our diaries for the last week in June.

Living close to the New Forest, Tim had developed a good working knowledge of the area, and despite having also placed his botany on the back burner over the intervening couple of decades due to a busy job and family commitments, his work contacts had nevertheless enabled him to carry out a bit of research work on the Forest's best floral locations prior to my arrival.

Designated in 2005, the New Forest is England's second newest and second smallest National Park, behind those of the South Downs and Norfolk Broads, respectively. Situated on the coast of central southern England, it has progressively become the last remaining substantial green space between the Solent to the east and Bournemouth to the west. The Forest is also a mosaic of habitats, which seamlessly blend into one another. Undoubtedly the most celebrated components are the largest area of semi-natural beech woodland in Britain and the most extensive remaining area of lowland heath in all of Europe – which is itself further subdivided into wet and dry heath. As a landscape it is also the last large relic of countryside from a bygone era, when vast tracts were considered as one grazing area unrestricted by fencing. First laid down in the Forest Charter of 1217, this covenant enshrined commoners' rights to graze their animals in this manner, and continues to this day, with ponies grazing on the heath, donkeys on the greens and pigs in the forest.

Botanically speaking, the New Forest only has one or two species of which it is the sole custodian (these being the wild gladiolus and possibly slender marsh bedstraw), so in this respect cannot compare to sites such as the Lizard or Breckland. But where the Forest excels is by being a stronghold for many native plants that have become rare or very scarce elsewhere. A further suite of plants exists within the Forest as a relict population, particularly in its bogs and wetter heaths, for example, that are traditionally considered more northerly species. Finally, a smaller group of plants in the Forest's woods and along its streams tend to be species considered to have a more westerly distribution.

After sending Tim my botanical wish list, he graciously declared he'd be more than happy to organise the day's events. His first job apparently had been to seek out the help of Clive Chatters – an acknowledged wildlife authority on the Forest. Armed with an array of locations containing the Forest's best plants, Tim had then simply linked up the dots that would see us travelling around the Forest in a roughly clockwise fashion. Meeting up just east of Ringwood, and with a recent change of legislation making Covid no longer quite the spanner in the works it had once been, we were able to car-share. Travelling together would not only help conserve fuel, but allow for some serious catching-up in between the bouts of hardcore botany. Being both Covid-free also enabled us to give each other a whopping big bear hug, something which neither of us had apparently done to folk outside our own immediate bubbles for quite some time.

Perhaps top of my botanical wish list had been a most peculiar member of the buttercup family. Looking more like a pocket plantain than the buttery-coloured denizens of a million meadows, I couldn't quite put my finger on precisely why I wanted to see mousetail so much. For starters it was a plant I'd never seen before, which will always lend a species a certain cachet, but perhaps even more importantly it was one of those plants that had the feel of the 'underdog'. The very specialised habitat of nutrient-rich soil that becomes flooded in winter and is frequently disturbed will always attract the kind of plant that prefers to hang out at the periphery of civilised botanical society.

Slotting straight into the groove like we'd only seen each other yesterday, rather than at least a decade ago, I suggested it might be easier if I drove while Tim took on the role of navigator – with the first destination being a disturbed trackway at a place called Tom's Farm, close to the western edge of the National Park's boundary. Clive had furnished Tim with details that upon arrival at the farm we should follow the public footpath cutting straight through the farmyard, before then looking at the poached (or livestock-trampled) areas to the left of the path. We'd also been relieved to hear that not only was the incumbent farmer friendly but also proud of being the custodian of one of the last and quite possibly largest population of mousetail still surviving in the Forest.

Reaching the spot to search, we could immediately see what appeared to be abundant suitable habitat, where the vegetation had been kept short by both farm traffic and cattle hooves, so split up to see who would find one first. Once again I had that slight niggling worry that we were approaching the end of the mousetail's flowering period. But with supposedly hundreds of plants here, surely it wouldn't be another plant intent on doing a 'spring gentian' on me? Largely unable to cope with pretty much any competition, the mousetail's survival strategy sees it capitalising on only the most ephemeral of locations, as and when conditions are suitable. But the only plant that I could immediately see that appeared to be colonising this precise niche was common knotgrass, which had spread its green net almost everywhere.

Tim's earnest, head-down demeanour suggested he hadn't located it either. So, after the best part of 20 minute's diligent searching, we came together for a chat as to whether we'd misunderstood Clive's precise directions. Further along the footpath to the left lay a cordoned-off pasture with a metal hay feeder as its centrepiece, which in turn was surrounded by a few grimy cows. Having been churned up by the cattle, the pasture looked more like a muddy scrap of the Somme than the National Park's finest remaining home to a rare and threatened New Forest plant. 'Surely the mousetail wouldn't be in there!' I murmured to Tim, who agreed, before then indicating that the terrain beyond looked even less favourable.

Hearing activity in the farmyard behind us, I suggested that maybe a better strategy would be to ask the friendly farmer if he could shed any more light on the mousetail's precise location. An exceptionally amiable character, he also knew the plant well, and after intimating we might already be a touch too late in the season, was also able to categorically confirm that the muddy area further along – which we'd just dismissed – was in fact the best area. The only problem that now quickly presented itself was making our excuses to get back to the site for another look, as our friendly farmer seemed eminently capable of talking the hind legs off a New Forest pony.

Politely taking our leave, we retraced our steps back to the field and launched ourselves into the mud. But at this location no plants immediately jumped out at us either, and after a further 15 minutes of searching, Tim suggested that with such a packed agenda ahead of us the smarter move might be to cut our losses and head to the next site. Hating to start with a failure, as it often seems to set a precedent, I persuaded him to give it five more minutes, and to be honest that was about as much squelching as we could hack anyway.

Then I heard a chuckle of delight. Turning around and in possibly the muddiest section of all, Tim was pointing down with that all-too-familiar twinkle in his eye.

Just one plant... We had no idea where all the other hundreds were supposed to be, but it was all I needed. Never has the phrase 'small, but perfectly formed' been so appropriate. After a quick congratulatory high-five with the plant finder I moved in for a closer look, which proved really difficult given that properly getting down to its level would have left me utterly filthy. It had to be the best plant I'd ever seen, in the worst location. The flower's columnar receptacle stuck up vertically like a miniature obelisk, and around its base were a row of upright stamens lined up like miniature matchsticks, which were themselves enclosed by the plant's five faint green and splayed sepals. The slightly fleshy leaves emanating from the plant's base were also decidedly grass-like in appearance. While trying to take a few half-decent pictures I felt a warm glow inside – we were finally up and running.

Briefly stopping off to pick up trailing St John's wort by the side of the road while crossing a site called Broomy Plain, our next official stop was Bolton's Bench. A well-known landmark within the Forest, Bolton's Bench is a gentle and heavily grazed hill, with a giant yew tree at its top that has been surrounded by a circular bench constructed to commemorate the eighteenth-century New Forest master keeper, the Duke of Bolton. The bench is also just five minutes' walk from the centre of Lyndhurst, the self-styled 'capital' of the Forest commandeered by William the Conqueror as the centre of his *Nova Foresta*. Most visitors to the attraction tend to take the short walk up to the tree for the view, but our focus instead would be a damp, hollow patch of lawn next to the offices of the district council.

Often growing on the hardest-grazed and most trampled parts of the Forest's lawns, our first quarry was that of chamomile, which Clive Chatters had labelled as a plant that 'has earned its place as the herb of humility'. Previously common on grasslands, sandy commons and damp woodland clearings across much of lowland England, it has since retreated to patches in southern England as a direct consequence of the decline in grazing. Chamomile thrives in those locations where the botanical competition is regularly cut back, and so can also be encountered on village greens and cricket pitches, for example, where the regular cycle of summer mowing and rolling mimics the actions of grazing livestock.

The heavily-nibbled New Forest is of course one of its strongholds, and sure enough we quickly located its thread-like, feathery leaves on a downward slope leading towards the spot where standing water was still present in a natural hollow. Right at the beginning of its floral season, one or two of its blooms were just at the point of opening, but unlike many wild plants where the flowers are frequently the key feature for clinching identification, in this case, verification is more easily obtained by your nose. Famously fragrant, chamomile's aromatic odour has been likened to anything from apples to bubblegum, but upon taking a whiff it immediately brought to mind my mother-in-law's herbal tea of choice.

But the damp patch close by held one more surprise, in the form of a very rare bedstraw that presented a trickier challenge to identify than that of the chamomile. Slender marsh bedstraw is virtually confined to the New Forest, with the only extant population outside the National Park being just across the county boundary in Wiltshire. But with both marsh and fen bedstraw looking exceedingly similar, would we even know it if we saw it? The devil, however, is in the detail, and upon finding a suitable candidate right on the edge of the water, Tim and I unholstered our eye lenses for a closer look. Supposedly smaller, slenderer and weaker than its two more-common counterparts, the plant's identification features only tend to work well when you are able to make a direct comparison, which we obviously weren't.

Having experienced fen bedstraw recently with the SRPG I was able to quickly eliminate this potentially confusable species from our thinking, due to the absence of a minute bristle at the tip of each leaf – leaving it as either marsh or slender marsh. Marsh bedstraw is commonly encountered across the United Kingdom in ditches, ponds, fens and marshes, and on the few occasions I'd given it anything more than the most cursory of looks, it had seemed more robust than the specimen we were currently poring over. The fruits also offer subtle but distinct differences between the two species, but it was way too early in the season to be able to use this feature. Supposedly the undersides of the flowers of slender bedstraw have a slightly more pinkish hue than those of its more widespread relative. But it was only after some seriously high-powered gazing that Tim and I eventually came to the conclusion that given the flowers' delicate rose-wash, the plant's scrawny nature and its precise location, then beyond reasonable doubt, it had to be 'the slender'.

Plants often refuse to fit into the precise pigeonholes into which we taxonomically oriented humans would like to place them. Many species, for example, exhibit huge natural variability, while others muddy the waters by hybridising with other closely-related species. And even in the company of botanical experts, I'd been initially surprised (and not a little relieved) to see them occasionally unsure as to exactly what they were looking at. The thousand-species challenge I'd set myself was something that could never be independently verified, and so by necessity would have to be self-ratified. This meant if I were to start cutting corners then I'd be only cheating myself. Here, honesty and propriety were to be the watch-words for my challenge with the bar set high. First and foremost I had to satisfy myself, and in the case of the slender marsh, the 'beyond reasonable doubt' was good enough for me.

Taking the B3056 out of Lyndhurst, Tim's next stop on our magical botanical tour was that of Matley Bog. An astonishing location situated right next to the road, Matley is essentially a stream-fed sphagnum-dominated quaking bog that slowly transitions further from the road into alder carr bog woodland. Necessitating wellington boots, this was a site where you could easily disappear up to your armpits in peat swamp if it wasn't treated with the greatest deal of respect. It is also a highly delicate site where carelessly placed boots can trample exactly what you're looking for, which in this case was bog orchid.

Making sure we avoided the bright-green sphagnum lawns, which can be the vegetative equivalent of quicksand, we carefully ventured onto the bog and were almost immediately rewarded with a fabulous array of plants.

Insectivorous plants are invariably confined to nutrient-poor habitats and in no time I managed to find both round-leaved and the wholly more unusual oblong-leaved sundew. Easily differentiated, the round-leaved's leaf-ends were shaped like a table-tennis bat, while those of the oblong-leaved were more reminiscent of a lacrosse stick. Possessing the characteristically sticky red hairs, these are employed by both species to ensnare any carelessly flying insects, which are ultimately converted into protein pills to supplement the plants' diet.

In amongst the marsh St John's wort, the first tentative flowerings of bog asphodel spikes also caught the eye. Possessing yellow star-like flowers on leafless stems, this plant really goes to town in the autumn by turning a tawny orange, giving the impression that large sections of bog appear on fire. Bog asphodel's scientific name of *Narthecium ossifragum* – or bone-breaker – also originates from the belief that grazing the plants would cause the bones of sheep to become brittle. We now know, however, that the plant was merely scapegoated, and it was the calcium-poor pastures in which the sheep had been forced to graze that were in fact to blame.

Calling me over with a tremulous voice, Tim had only gone and done it again. Often considered a holy grail plant, bog orchid is not only our smallest British orchid but is also considered one of the trickiest to track down. Thriving only in the geographically distanced locations of the New Forest and the western Scottish Highlands, it is not only scarcely distributed, but with its predilection for embedding into similarly-coloured sphagnum, picking out the flowering spikes requires a good and experienced eye. Which is exactly why Tim didn't want to look away from the precise point where he'd just spotted one until I clapped eyes on it too.

As I squelched over, the object of Tim's gaze could in essence be described as dinky and exquisite in equal measure. Only marginally taller than the length of my index finger, the spike of minute pale green flowers could clearly be seen emerging from a pair of elliptically shaped basal leaves. I had previously seen the orchid both up in the peat bogs of North Wales and on Fetlar in Shetland, and so completing the Triple Crown wasn't quite enough to make me sink into the peaty goo in celebration, but it was another pleasing find. Being especially careful not to crush this most delicate of plants to death, we quickly scanned the surrounding vegetation, eventually finding a grand total of five spikes before retiring to the car to dry off down below and re-apply sun lotion up above.

Having flowered far earlier in the year, the native *Pulmonaria* that is narrow-leaved lungwort was next on the itinerary. But as all that currently remained of this rarity was a pile of spotty leaves, if I were to be honest the plant now represented little more than a botanical tick along a roadside verge. As our earlier mousetail endeavours had now left us well behind schedule, and with bigger fish to fry elsewhere, I'm afraid to admit it was given nothing more than a cursory glance as we headed straight off for a date with a pond.

The plan was to head south to the tiny hamlet of East End, where an eponymous freshwater pond would hopefully offer us a different suite of species to those we'd observed so far. Like Matley Bog, the pond itself was right alongside the road, which meant that having safely pulled off the road our commute to work consisted of little more than just a few steps. Ponds are wonderful in that they tend to offer two habitats in one: a marshy fringe and the open water itself. Dividing to conquer, I tackled what lay in the water while Tim worked the margins, and within seconds we'd both found a cracking plant each.

Having instantly latched onto one of my target plants, I was reminded of a casual remark made by my friend Justin, the steel erector, whose acquaintance I'd made earlier in the year when tracking down Bristol rock-cress. His observation that plants named after locations must be rare rang true for the reddish-green aquatic plant that I'd quickly spotted poking out of the centre of the pond. Although not entirely confined to the county, the vast majority of Hampshire purslane's records do fall within the county's boundaries. Hampshire purslane, although admittedly not spectacular, is one of those plants that all visiting botanists need to see, and – thanks to both its colouration and ability to quickly proliferate at certain locations – is also one of the easiest to find. Most of the leaves below the water surface were still largely green, while only those exposed to the sun and air had begun to turn their characteristically deep purplish-red. Apparently as the summer proceeds, and presumably helped by dropping water levels, virtually the whole plant responds by turning a dark red. So eye-catching are these displays that vigorous populations can even be spotted from a moving car once you know what you're looking for. But while seemingly in very rude health at East End Pond, the purslane must still be considered a rarity, with southern England very much at the edge of its climatic range.

While pootling around the pond's margins, Tim had located the second plant that I'd never seen before in as many minutes. Pennyroyal is amongst those plants that can be placed into the same cadre as that of Hampshire purslane – common in the Forest, while difficult to catch up with elsewhere. Considered to be a native plant in the United Kingdom, the issue of its natural distribution has been clouded by accidental introductions elsewhere, but with the Forest both one of its major strongholds and at the northern edge of the species' range, few would argue that this was also the best place to see it. A pungent member of the mint family, its whorls of pinkish-purple flowers were unfortunately not yet in evidence. But perhaps pennyroyal's most eye-catching trait was its prostrate and creeping growth form, which must be considered unusual within the context of the rest of the family, which are largely upstanding members of the community. Considered something of a panacea in herbal folklore, the plant certainly possesses sedative properties and has even been used as an abortifacient. But as the active ingredient pulegone is capable of causing liver damage, this is one mint that shouldn't be mixed up with its decidedly more benign relative – spearmint – when making a brew.

And the fun didn't stop there … By working the margins, Tim and I quickly located yet another plant that had patently adopted pennyroyal's habit of flying (or creeping) under the radar. Coral necklace is most commonly encountered close to temporary pools in heathland or heathy grassland, and in seasonally flooded hollows along sandy or gravelly tracks. So-named because of the clusters of pinkish-white flowers, which appear strung along the trailing reddish stems rather like pearls on a necklace, all we could see on our visit was the unadorned necklace, as its 'pearls' were not slated to appear until July at the earliest.

Coral necklace is a plant with an interesting distribution. First identified in 1666 when specimens were sent (once again) to the celebrated botanist John Ray, the first New Forest records, by contrast, only date from 1925. There has been speculation that the coral necklace here may have been imported to the New Forest with some nursery stock of young conifers, although this remains unproven. What is indisputable is how much the plant seems to have found the landscape of the New Forest to its liking, resulting in its spread across the National Park and even over into Dorset and the Thames basin. This is quite the opposite with how the plant appears to have fared back in Cornwall, where a serious decline has been charted for close to a hundred years. But in a dramatic twist, the plant was very recently and sensationally discovered growing in abundance close to a remote loch in the Scottish Highlands. Hundreds of miles north of what was previously considered its natural distribution, this discovery perhaps only re-emphasises that the more we think we've learned about plants, the more we realise we know virtually nothing about them at all.

After our slow start we were now undoubtedly on a roll. And this luck continued when successfully catching up with the diminutive and frankly splendid yellow centaury. Only commonly encountered in the Forest and on the Lizard, where it can be seen sticking its (flower) head no more than 5 or 6cm above the Forest's grazed lawns, this represented my 11th lifer of the day – the National Park and my old pal were delivering in spade-loads.

We were now on the home straight, with just a few more stops planned before I dropped Tim back at his car and we headed our separate ways. On my hit list of New Forest targets, an intriguingly-named plant called bastard balm was another species I was really keen to have a crack at. The origin of its rather abusive name appears to pre-date the current meaning of 'bastard', and may hark back to the late fourteenth century, where the word 'bastard' was used to describe something 'spurious, or not genuine, but having the appearance of being genuine'. This enabled Medieval botanists to distinguish it from the 'true' or lemon balm, which is ironic when you consider that out of the two, only 'the bastard' is considered to be a native British species.

Along with slightly disturbed paths and hedgebanks in South Devon and Cornwall, the New Forest was considered the balm's only other stronghold – until recently that is. Probably due to both a lack of coppicing and systematic overgrazing, the plant has declined in virtually all of its open woodland sites, with Roydon Woods Nature Reserve now considered the only place within the Forest where a stable population can be encountered. Managed by the Hampshire and Isle of Wight Wildlife Trust, the reserve is close to 400 hectares in size, so simply wandering in to look for the plant from scratch would have been nigh on impossible given our time available. But thanks to some more invaluable information from the incomparable Mr Chatters, and a helpful contact of Tim's at the Trust, he'd been able to secure some VIP access into the remotest part of the wood, along a farm track and through a couple of locked gates.

Having been baked in the sun all day, the cool shade of the woodland canopy offered us welcome respite as we entered what appeared to be a forester's trail, well away from any of the reserve's other paths. The reserve is convoluted in shape, and bisected north to south by the meandering River Lymington. Mostly comprising wooded blocks, where predominantly beech, oak and hazel were mixed in amongst planted conifers, it also includes a number of open pastures, which, despite being mostly within the wood, appeared also to be outside the Trust's remit.

Due to the time of year, the wood had that 'twixt cup and lip' feel about it – too late for the spring woodland flora, yet too early for the fungi, which would begin appearing as high summer slowly morphed into early autumn. July was almost upon us too, traditionally a quiet month for birds as many are discreetly raising broods, while others are preparing to moult and/or migrate. Supposedly an overly large population of introduced Sika deer resided within the woodland too, which we could only presume must have been resting up in the more secluded parts of the forest before waiting to emerge at dusk.

Clive had advised us that perpendicular to the forester's trail ran an old wooded bank, and if we were to follow this then we might have a chance of encountering a balm or two where some recent coppicing had increased the light levels on the forest floor. Looking like a gigantic mole had been busily at work down below, the distinctive bank was soon located, and sure enough in an area a little less shady than its surroundings, the plant was patiently waiting for us. Calling Tim, who'd had the misfortune of having chosen the other side of the bank to scour, I basked in the glory of being the one to find it for just a few seconds longer than was commensurate with my age, before then getting down to the serious business of enjoying it. Much larger than I'd originally thought, it was way more attractive too, and Tim shared in the delight that occurs when a plant totally surpasses your expectations.

Slightly frustrated that we'd been either too early or too late in the season to see a number of plants at their best during the day, this definitely wasn't the case here, as the blooms looked immaculate. The large, zygomorphic (one plane of symmetry) flowers were mostly white but with a bright fuchsia-pink lip, and if we had been able to admire them in ultraviolet light – as pollinating bees do – then presumably they would have stood out even more. Apart from the sumptuous flowers, the presence of hairy, toothed pairs of leaves placed at regular intervals up an equally hairy stem seemed much more representative of your average labiate. But taken as a whole there was nothing average about this plant, and it was in fact the 'bastard' that was the genuine article, not its lemon-scented imposter.

Finding such a fabulous plant was the botanical equivalent of downing a double-espresso as we dashed onward with renewed vigour to our last destination, stopping only briefly to catch up with our third and final sundew of the day, in the form of 'the great'. Exceptionally rare away from the Highlands of north-west Scotland, the presence of all three sundews so close to the south coast only served to re-emphasise what I now knew to be categorically the case with the New Forest – that it really is very good for wild plants.

Finishing more with the flair of an impresario than that of a conservationist, Tim had arguably saved the Forest's most special plant for last. Now heading to the large forested inclosures north of Burley, we were off on a wild gladiolus hunt. First discovered in the Forest in 1856 by the Reverend W. H. Lucas, its discovery here was somewhat wonderfully pipped by a record from the Isle of Wight by a Mrs Phillipps from the previous year. The gladioli on the Isle of Wight were unfortunately not seen again beyond the 1930s as their former localities were converted to a golf course and turned into quarries. The remaining wild gladioli in the Forest have also been subjected to some rough treatment through the ages, but healthy populations do still exist – if you know where to look that is.

Having been shown wild gladiolus in the Forest around 20 years previously, the couple of facts I could remember about where to look for them were that the woodland edges or rides tended to be the most productive hunting grounds and that they liked hiding beneath bracken. Clive Chatters remarked that the gladiolus 'is not so much rare, as elusive', with the 1996 *Flora of Hampshire* showing the plant to be widespread in the southern and central parts of the Forest. Despite this, the fear of the sleek and beautiful flowers being selfishly picked or carelessly trampled remains to the extent that the exact locations of gladiolus stands are often jealously guarded. However, armed with some privileged information furnished by none other than Clive himself, Tim and I felt we'd again been granted VIP access to a most wonderfully exclusive club as we marched off across a grassy heath in search of woodland edge, bracken and what might lie beneath.

To help when looking for the gladiolus, it can be instructive to learn about the management of the Forest over the last 170-odd years. Certainly the Deer Removal Act of 1851 had far-reaching consequences for the history of the Forest, which continue today. Under this Act of Parliament the Crown surrendered their right to keep deer, but as compensation then took the power to fence and plant thousands of acres of the open forest with timber trees. These forested areas became known as inclosures.

As deer numbers plummeted, the inclosures experienced a period when grazing pressure was reduced, and as the newly planted timber had not yet managed to shade out the forest floor, this favoured the gladioli. But ultimately this did not last, and they were crowded out by the conifer stands. However, research revealed that the gladioli also seemed to favour the richer, drier soils where the woodland grades into grassy heath. Here the continued presence of the trees close by helps with both the enrichment of the soil, and by preventing the grassy areas from developing a covering of heather. In addition, these areas tend to be where bracken flourishes, which helps hide the slightly shorter, and somewhat tastier, gladioli from grazing animals.

Arriving at a large inclosure with extensive stands of bracken alongside, Tim and I intuitively knew we'd arrived at the place close to where the 'X' would have been marked, were we following a treasure – rather than an OS – map. Splitting up to cover a larger area, we steadily began lifting fronds to see what lay below. The fine art of 'bracken-lifting' was quickly learnt by the rush of visiting botanists following the gladiolus' discovery, with Dr J. T. Boswell Syme writing extensively on the precise technique as early as 1863.

So it only felt fitting and proper that Tim and I followed in the grand tradition. No, no, no... a few more paces, no, no, no... a few more paces, no, no, oh! The shocking, gaudy colour of its flowers were physically and metaphorically breathtaking. Looking over to Tim, I could see he was waving to me as I yelled across in a state of delirium, 'Have you got one too?' 'Loads!' came the gleeful response. Timing-wise we'd nailed it once again, as I counted five flowers, all at slightly different stages of emergence. The bottom two had completely opened to reveal a much paler pastel pink on the inside of each petal, while the uppermost flower was just beginning to emerge from its purple-tipped bract. Each bloom additionally had a curved perianth tube, which gave the flowers the impression of being delicately painted long claws.

Locating another couple nearby, I went across to join Tim, who'd seemingly landed the mother lode with five perfect specimens all within touching distance of his feet, which were firmly planted in a layer of bracken litter. Filled with botanical glee, Tim suggested we repeat this exercise every year, insisting we get our diaries out for 2022, when next time we should instead head out to chalk downland. And that, in a nutshell, is another reason why plants are so wonderful: they possess the power to reunite old friends.

CHAPTER 11

Back to Uni

Having taken on the challenge of seeing a thousand different British plants in a single calendar year, I knew a trip to North Wales would be a must for a number of reasons. Firstly, having lived up in North Wales for the best part of five years, while working for the North Wales Wildlife Trust and studying for my master's degree in ecology, I had developed a good working knowledge of many of its best wildlife sites. Secondly, North Wales encompasses everything from the mountainous regions of Snowdonia to the base-rich fens on the Isle of Anglesey, enabling it to punch well above its botanical weight. And perhaps most importantly of all, it also served as the base of a certain Nigel Brown.

When asked who has been the 'wild' inspiration and driving force behind my career, many expect me to mention one of the two Davids: Attenborough or Bellamy. While it's true that I've derived huge enjoyment from watching them champion the natural world in their very different ways during my formative years, neither, in my humble opinion, come anywhere close to Nigel Brown, my old university lecturer. Now enjoying his retirement, Nigel was a constant, reassuring presence for the entire length of my tenure up in North Wales, in his dual roles as biology lecturer at University College North Wales (otherwise known as Bangor Uni) and curator of the university's botanical garden at Treborth.

I first met Nigel back in 1991, when I was working as seasonal warden at the Cemlyn Bay Nature Reserve. Managed by the North Wales Wildlife Trust, this small reserve on Anglesey's northern coastline looks out into the Irish Sea and stands in the shadow of Wylfa Nuclear Power Station. In summer, the reserve's brackish lagoon and low-lying islands, which are protected from the full force of the waves and tides by a large shingle spit, become transformed into a seething and vibrant tern colony. Here common, arctic, sandwich and just occasionally roseate terns, nest cheek by jowl, and my job as warden was to both protect the birds and welcome the human visitors coming to enjoy this most delightful of summer spectacles.

Nigel had been one of those visitors I'd been tasked to meet and greet, and I remember being immediately impressed with his utterly charming demeanour and seemingly encyclopaedic knowledge of both the reserve and the wildlife across Anglesey. Kindly advising me where else I might visit on the island to watch wildlife in my spare time, he also invited me to drop in at Treborth for a personal tour at any point. Fast forward a few months, and having finished my wardening contract, I'd decided to stick around in the small Anglesey town of Menai Bridge, named after its famous suspension bridge, which links Anglesey to the mainland. My plan had been to use my very first paid job in conservation to help wheedle out other gigs, with the aim of ultimately stitching them together into a fully-formed career.

With Treborth Botanic Gardens situated just across the Straits and right next to the mainland end of the bridge, it didn't take long for me to follow up on Nigel's kind offer. Touring with him around the gardens, I was astounded at his ability to one minute be waxing lyrical on lichens, before then espousing the beauty of ferns or fungi – all it should be said with the most remarkable modesty; for Nigel sharing was caring. Finally, before I cycled back to my digs, he asked me if I'd like to look at the moths that had been caught in the light trap the previous evening. Sitting down with a cuppa, he then proceeded to open my eyes to the wonderful world of Macrolepidoptera, as I took in everything from peppered moth to peach blossom. Running right through the year, the moth trap was systematically emptied each morning, and I quickly sensed an opportunity for the creation of a job as Nigel's 'Assistant Moth-er'. Immediately hooked, I then returned for some more moth action the following day, with this activity quickly slotting into my daily routine in between showering and taking breakfast.

Each morning upon my arrival, Nigel – who lived on site with his family – would be waiting with a cup of coffee and a ready smile as we chatted moths and so much more. Scarcely missing a session for well over two years, I came to cherish the time spent in his fine company. I also quickly realised that much like the popular American sitcom *Everybody Loves Raymond*, everyone indeed loved Nigel too. He was simply adored by students and colleagues alike, while those in the conservation and wildlife-watching community regarded him with the utmost reverence.

When I went on to study for my master's thesis over the bridge in Bangor, my friendship with Nigel, his wife Caroline and their two terrific kids strengthened, and I saw my relationship change from that of young apprentice to family friend. Branching out from the daily mothing sessions, I also became a regular participant on Nigel's botanical field trips as we found ferns and searched for sedges. Fast forward 30 years and we've remained in contact, and our respective families seize the opportunity to meet up whenever possible.

A monumental effort by Tim in the New Forest the previous day had seen us achieve a clean sweep of all the species I'd hoped to see. And as I travelled up to where Nigel and Caroline now lived on Anglesey, my fervent hope was that with my old lecturer's help, my one-thousand target would finally begin to look a little more achievable.

Arriving just in time for dinner at their fabulous cottage in the heart of Anglesey, I hoped the warm welcome I received would then fire the starting gun on two whole days of plant-filled fun. Having not caught up with the Browns since BC (Before Covid), there was much to discuss, as I found out how their kids Daniel and Laura were faring, while they were keen to hear news on Zachary, Christina and how my thousand-plant challenge was going. And while Caroline applied the finishing touches to dinner, there was still time for the equivalent of a 'botanical starter', as Nigel had a surprise up his sleeve along the roadside verge just outside the cottage.

Having been part of a team that encouraged Ynys Môn (Anglesey) Council to practise a plant-friendly mowing regime along many of the island's flower-rich roadsides, Nigel had subsequently been rewarded with the appearance of a few frog orchids just beyond his doorstep. Surely a candidate for the title of 'Britain's fastest declining wildflower', the frog orchid was another of those plants more suited to an era before huge swathes of countryside became 'agriculturally improved'. More recently, any surviving populations have suffered another setback due to a lack of grazing, leading to this grassland specialist being choked out as the rank species take over.

And the particular specimen he showed me was a leviathan. Rising from its basal rosette, the orchid spike was also beautifully lit by a shaft of golden evening sunlight, with the effect of placing it front and centre. Frog orchid is a species capable of growing even in the arctic tundra of Shetland's Keen of Hamar, Britain's most northerly reserve for orchids, where the tough conditions see it reaching little more than 5 or 6cm in stature, but in the benign climate of an Anglesey roadside verge, the illuminated specimen had reached a height close to that of my son's 30cm school ruler. It has to be admitted that turning the flowers into frogs requires a feat of imagination almost as challenging as that of turning frogs into princes. But if you squint, then the flower's pallid green lip can take on the appearance of a frog's hind legs in mid-jump, while the helmet of sepals *almost* form a frog's squat body with the forelimbs tucked neatly under its chin. Although definitely one of our more understated orchids, what could not be overstated was the sheer number of flowers on this particular stem. Quickly totting up a grand total of 37 frogs on the single spike alone, I mused that the only element seemingly missing from the scene was a dollop of frogspawn nearby.

Joining the Browns for breakfast after a very peaceful night's sleep, I was delighted to have a coffee pushed into my hand as I properly took in their garden for the first time. Nigel and Caroline had found that the peace and tranquillity brought about by the pandemic had been one of its very few bright spots. Trying to use the enforced period at home as effectively as possible, they'd devoted themselves to their garden, and I had to concede it had been time well spent. With their property attractively positioned at the base of a small quarry, they'd managed to make the most of the topographical features by creating what was in effect a (nearly) secret garden. With a variety of habitats packed into an area marginally larger than a tennis court, the trees, shrubs, flower beds and water features all appeared to seamlessly flow into one another. Sections of the lawn had been left uncut to create miniature meadows full of cowslips, and taken as a whole, their green plot looked every inch the poster child for how every wildlife garden should look.

Having discussed a strategy beforehand, we'd decided that covering all of North Wales' main locations would have been simply impossible. A trip to the Scottish Highlands would follow directly on the heels of this trip, giving me ample opportunity to catch up with a variety of montane species. So on this occasion, the smart move would be to give Snowdonia a miss. Our plan would therefore focus on

Anglesey for the first and only full day, while saving the huge limestone lump of the Great Orme back on the mainland for the day after. With the Great Orme located back along the North Wales coast and closer to the English border, this meant that as Nigel headed back home to Anglesey, I'd simply continue towards the M6, thereby giving me a head start for my journey northwards up to Scotland.

While Nigel talked through our plan of attack, my attention was briefly drawn just beyond the patio doors once more, not by the plants this time, but the sudden guest appearance of a couple of red squirrels attracted to feeders placed at the edge of the terrace. Following a successful project to eradicate the non-native grey squirrels from the island, and the bolstering of the indigenous red squirrel population from captive-bred stock, Anglesey has now become a haven for a species that I certainly never saw when I lived there. Sometimes all is not quite as doom and gloom on these crowded little islands as we like to think, I mused as I turned my mind back to the more pressing matter – that of ensuring our day would be packed with plants.

With our botanical paraphernalia duly stowed in Nigel's car, we then headed for one of Anglesey's most iconic and well-visited reserves. South Stack RSPB Reserve is situated on the diminutive Holy Island just off Anglesey's west coast, and easily reached via a short road bridge. As the reserve is mostly composed of farmland and heathland which sit above a stretch of dramatic sea cliffs, the vast majority of visitors at this time of year are after views of choughs and the thronging seabird colonies strung out along the cliffs' impossibly narrow ledges. Not us though, as one of South Stack's best-kept secrets is its suite of rare coastal plants, which are frequently overlooked by the visitors in their dash for some puffin therapy over at the cliffs.

Arriving at the reserve's car park, I realised just how long it had been since my last visit as I took in the new visitor centre and cafe that had sprung up in my absence. Hitting the ground running, we worked the margins of the car park first, with the stout, woolly white stem of great mullein and the spotty stems of hemlock the first of the day's 'year ticks'. Notoriously poisonous, hemlock has gained infamy as the plant given to Greek philosopher Socrates at his execution. By all accounts it would not have been a pleasant death, as paralysis, respiratory failure and stupor all kick in before a loss of consciousness finally marks the beginning of the end. Mercifully, the repugnant smell of the crushed leaves tends to give you more than enough advance warning that the foliage of this 'parsley lookalike' is best left well alone.

Aware that we had little time to dawdle, we set off along the footpath that would take us past South Stack's seabird cliffs and ultimately up onto the heathland's more exposed regions, where a selection of botanical delights, we trusted, would be waiting to welcome us. The path-sides along the way were a riot of colour, with wild carrot, bell heather, tormentil, bird's-foot trefoil and spring squill all jostling for position. And despite these species having previously been logged elsewhere, they were nevertheless very useful in helping me warm up my plant skills by reeling off their names as we went.

Alongside the three heather species, the only other dominant plant across the heathland was western gorse. With a much more westerly distribution than that of its far more widespread cousin, the European gorse, western gorse has a distinctively shorter flowering season. Known for its egg-yolk-yellow flowers being visible throughout the year, this is perhaps why the saying 'When gorse is out of bloom, kissing is out of season,' was originally coined. However, if the same saying were applied to western gorse, then the kissing season would be disappointingly truncated to little more than July and August. Visiting as we were on the last day of June, it was perhaps fitting that the very first flowers had only just begun to unfurl from amongst the gorse's decidedly unpalatable foliage.

Higher up on the headland, and as the heath became more open to the elements, we could see conditions turning increasingly inhospitable. Here the terrain was not only frequently subjected to long summer droughts, but also had to endure an almost incessant battering from salt-laden winds, leaving only the toughest plants able to eke out a living. Looking more like barren arctic tundra than lowland heath, the stony plateau which represented our destination was nowhere near as botanically diverse as the sheltered paths below, but paradoxically the plants present were far more interesting.

Nigel, in his capacity as the BSBI recorder for Anglesey, had the distinct advantage of knowing exactly what to look for and where to find it. Joining him on his hands and knees, I could see at least half a dozen perfect and perky little pink flowers delicately studding the stony ground all around us. Reaching a lofty height of no more than 6cm above the terrain at best, and to which they were anchored with the help of a small basal rosette, I instantly recognised the blooms by shape as belonging to a group of plants called centauries. Identification to species level would need a quick consultation from my BFF, but having forgotten to inform Nigel that I usually prefer to try working out for myself exactly what I'm looking at, he quickly gave away its identity as 'seaside' before I'd reached C in the index.

'That's a new plant for me, and my fourth centaury of the year!' I breathlessly exclaimed, after a quick confirmation from my book. Being a sub-set of the gentian family, there are only seven different species of centaury (and a hybrid) listed on the British flora, and the fact that I'd already logged four of them was perhaps the clearest indication yet that despite still finding myself well short of my ultimate target, it was also good to remind myself how many plants I'd already seen.

Arguably the two greatest thrills – and challenges – in field botany are firstly finding the more unusual plants and secondly managing to identify them. Certain plants, such as sand crocus and mousetail, can be very difficult to track down, but are a piece of cake to recognise, while for other plants, such as slender marsh bedstraw, finding them is much more straightforward than correctly confirming their identity. Some plants, of course, are both taxing to find *and* identify!

For me personally, grasses have always come into the 'difficult to identify' category. As a group they require time, effort and an in-depth knowledge of botanical terms, meaning that I'd progressed little beyond learning around a dozen of the most cosmopolitan and distinct species. I was also aware that this wanton disregard for certain difficult groups would need to change during my challenge.

The unimaginatively, but accurately-named heath grass could perhaps be described as 'unremarkable' in the looks department. Possessing anywhere from three to twelve ovoid to oblong spikelets in a narrow, compact panicle, the grass also possesses a somewhat obscure fringe of short hairs at the stem's ligule. In short, its most distinctive feature is not having any distinctive features. So it was a thrilling moment when I subsequently managed to instantly identify a species of grass close to the centauries, which up to that point I'd never knowingly even seen before. Patently the hours spent poring through the field guides on any number of evenings was finally and gloriously starting to pay off.

Elsewhere on the rocky terrain, we also managed to find goldenrod, which was a plant I was far more used to seeing at lofty elevations in the Scottish Highlands. The plant's altitude on South Stack's sea cliffs must have been no higher than 125m above sea level, and it was astonishing to think that close to the summit of Cairngorm the very same species could also be found flourishing at over 1,000m. But despite the very obvious climatic differences between both locations, one feature they shared was how few other species they had chosen to live alongside. Plants that are capable of living on the edge, it would seem, can take all manner of weather extremes in their stride, and the only occasions when they begin to struggle is in the face of botanical competition from other species. In other words, if you can conquer the climate then you may well have the terrain (almost) to yourself.

The only other key species up on the rocky heathland that Nigel wanted to track down was one of our rarest rockroses. It is quite possibly a toss-up as to whether white or spotted rockrose takes the title of 'Britain's rarest rockrose'. Both species are confined to just a few localities in Britain, with white rockrose known from only four sites in Somerset and four in Devon, while its spotty counterpart has been noted at only six sites, either on Anglesey or North Wales' Lleyn Peninsula. However, a visit to either Somerset's Brean Down or Devon's Berry Head in early summer will reveal the white rockrose to be in such abundance that it can sometimes look like these sites have suddenly been subjected to an unseasonal snow flurry. The same cannot be said of spotted rockrose, however, as its flowering plants are vastly more difficult to locate.

To make matters even trickier, the few plants of spotted rockrose that do decide to flower will often end up dropping their petals by midday, and may not even bother opening at all on days when the weather is inclement. Another rocky terrain specialist, the plant's downy, three-veined leaves were soon located, but with the weather both windier and colder than we suspected the plant would appreciate, it did not look terribly hopeful for flower-spotting. However, being always up for a challenge, and after 15 minutes of a painstaking and knee-bruising search we eventually managed to find a single plant that had decided to stick not one, but two flowers above the parapet. The flowers' five petals were cowslip yellow in colour, with each leading edge crenelated like a child's milk tooth. But easily the most distinctive feature was the mauve-coloured smudge (or spot) at the base of each petal. Glancing at the time, I realised these subtle and gorgeous petals would probably only remain attached for another couple of hours, before then being left to the mercy of the wind.

When living up in Anglesey this was a plant that I'd successfully seen before with my good friend Tim Sykes, but having started our trip out too late on that occasion, by the time we'd finally managed to find the plants, most of the flowers' petals had dropped. It would seem the early bird not only catches the worm, but also gets to see one of Anglesey's rarest plants with its petals still attached, I mused, as we headed back down to the cliffs.

Casually picking up (for me) another lifer, in the form of hay-scented buckler fern tucked into a gully along the cliffs, we then headed to the classic viewing point of South Stack's seabird colonies right next to the iconic Ellin's Tower. While most people spend their time here scouring the cliffs for puffins across the gully, mine and Nigel's job was instead to scan the short maritime turf just in front of our noses that covered the clifftops. *Tephroseris integrifolia* ssp. *maritima* is a plant that is found nowhere else in the world. More commonly known as South Stack fleawort, it is considered as a subspecies of field fleawort, which in turn is an uncommon plant of south-facing and thin-soiled chalky grasslands in southern England.

We jostled for position at the viewpoint with a number of birders; some even looked vaguely amused upon learning that Nigel and I were primarily 'here for the plants'. The fleawort itself was easy enough to find, but being located on a 45-degree slope, and with the vertical drop-off down to the sea just beyond, getting any closer would have been complete madness. Although I could see that it superficially resembled a ragwort, with between two and five flowers atop a down-covered stem, I decided I wouldn't push my luck by inching any closer, as *593 Shades of Green* didn't quite have the same ring to it. On just this one occasion, discretion, and a distance of around 5m, would be the better part of valour.

Sand dunes and their associated communities are undoubtedly one of Anglesey's most important habitats, and so were a shoe-in for a visit from Dilger and Brown. The prevailing winds and metronomic tides have brought sand ashore here for thousands of years, which in turn has created huge, dynamic sand-dune systems that pepper the south-west coast of the island. Perhaps the best-known example is that of Newborough Warren at Anglesey's southern tip. It is not only one of the largest and finest dune systems in Britain, but in recognition of its outstanding importance for wildlife it was also designated as Wales' first coastal National Nature Reserve (NNR) as far back as 1955.

While the NNR would represent our last stop of the day, Nigel first wanted to visit the dune system just a touch further along the coast at Aberffraw, as its smaller, more-compact nature would make both accessing and finding plants far easier than at Newborough. And pulling into Aberffraw's car park, we could immediately see the whole dune complex laid out in front of us like an enormous undulating and rippling carpet.

Most sand-dune systems tend to be characterised by ridges or hillocks of sand that initially form just beyond the reach of the highest tides, before ultimately running in bands parallel to the sea. As each line of dunes ages it also becomes progressively taller and more vegetated while the formation of new, embryonic dunes on the seaward side effectively pushes the older dunes ever further inland. But these dune ridges also frequently become interspersed with linear low-lying depressions, called slacks. Here the terrain is flatter, wetter and greener than the surrounding sand dunes, and the more amenable conditions in these slacks frequently house a higher diversity of plants than on the perennially windy and desiccating sand dunes just above.

As we marched out to one such slack, the immense variety of plants quickly became apparent as a drift of marsh helleborines competed for my attention with the far less numerous, but far more obvious, specimens of early marsh orchid. While I'd successfully managed to log both of these orchids elsewhere on my travels earlier in the season, the bright red form of the early marsh orchid (or subspecies *pulchella*) was quite breathtaking. A member of the genus *Dactylorhiza*, early marsh is just one of a group of orchids that while being closely related can also be individually quite variable. To make matters worse, hybridisation is relatively common within this genus, causing even seasoned orchid experts to disagree as to which populations should be classified into species, subspecies and varieties. Personally I've always been more in favour of 'lumping' than 'splitting' and having already recorded 'early marsh' in Upper Teesdale's pastures was good enough for me.

Re-reminding myself that there is so much more to botany than just looking at orchids, I began casting around for the arguably less glamorous species, but plants that were every bit as interesting ... and perhaps even more importantly, plants I still hadn't seen. Ever since my visit to the Lizard I'd gained a whole new appreciation of the wonderful world of clovers, to such an extent that the second I spotted a specimen close by, I knew it to be one I hadn't seen before.

The beautiful pompom-like pink flower heads were on the ends of long stalks, which rose above a bed of trifoliate leaflets, all of which had a weird venation that saw the veins recurved towards the leaf margins. 'Is this strawberry clover?' I asked Nigel, who confirmed my suspicions in a manner that conveyed both encouragement and congratulations. I was thrilled with the find, as I was pretty sure that before starting my big botanical year I'd probably only seen two species of clover – the omnipresent red and white – but following a quick consultation of my BFF I reckoned I'd now seen a mightily impressive twelve. But with so many other plants to check out, there could be little time to waste by indulgently counting clovers.

I'd also realised early on that the key to securing a high and handsome total would be finding and successfully identifying those plants that were neither fantastically rare nor ridiculously common, but somewhere in that huge middle ground. A perfect example of this latter category was when Nigel pointed out corn sow-thistle to me. Both smooth and rough sow-thistle were plants that I'd already seen a hundred times each during the course of the year, but their 'corny cousin' had not even figured on my radar. On proper inspection it really was quite distinctive too. Its deep, yellow flowers were much larger and quite a bit showier than those of the other two closely related species, but perhaps its most memorable features were the sticky yellow hairs that covered the bracts just below the flower heads. It really was a classic case of not judging a plant by its covers.

Nigel was now in his element as he pointed out a trio of grasses that I was sure I'd never seen before. Squirrel-tail fescue, sand cat's-tail and early sand-grass were all dune specialists, and in the case of the sand-grass he'd unearthed a real rarity too. Found only on bare, sandy ground by the sea, it is believed that the only place in Britain where this diminutive grass grows naturally is Anglesey. It is also a grass that flowers so early in the year, that the species now looked utterly frazzled by the time of our visit. With me incredulous that he could even identify the grass at all given its advanced state of desiccation, Nigel went on to explain that even when the plant was well past its sell-by date it always seemed to retain a peculiar 'jizz', with its stems and leaves resembling a pin cushion. Not for the last time that day I felt immense gratitude that Nigel had been so ready to help me with my challenge.

And the new plants kept coming. The mouse-smelling houndstongue, blue fleabane and round-leaved wintergreen were all clocked in double-quick time, with Nigel constantly asking whether I 'needed' each species for my list, while pointing out a succession of amazing plants. When initially devising the competition rules for my year I'd decided early on that grasses, sedges, rushes, ferns and trees would all be included, as 'a thousand different flowers' would have simply been impossible given work and family commitments. So when Nigel pointed out an interesting horsetail, I was all ears. Horsetails often get a bad rep, certainly from gardeners due to field horsetail's (or mare's tails') invasive nature, which can see it quickly taking over a flower bed with the help of its underground rhizomes. Difficult to evict once it has taken hold, I have clear memories of growing up with my father in a constant war of attrition with what he deemed to be the 'pernicious plant' in his beloved rose beds.

But this ancient lineage of plants that date right back to the Jurassic period is about much more than just one species, as the genus *Equisetum* encompasses a number of very attractive species, living in specific habitats and with restricted distributions. Pointing out one such example poking no more than 7 or 8cm out of the dune slack, Nigel declared that I'd definitely need this one. Turning a burnt orange, and with each delicately jointed stem topped by a cone that looked like the world's smallest pineapple, variegated horsetail instantly claimed the title of 'plant of the day'. Nigel went on to explain that variegated horsetail has a peculiar distribution, which sees it ranging from sand-dune slacks at sea level (such as here at Aberffraw) to mountain ledges and stream-sides. This was a plant that was equally happy at both sea level and 1,000m above sea level, but abundant at neither.

Retiring to my hosts' home for a short break from botany and an early dinner, Caroline then suggested, much to our delight, that she'd like to join us for our last site visit of the day: a walk along Newborough Warren's beach to the adjoining Llanddwyn Island. With Nigel and I having struck lucky with the weather all day, the evening looked to be a continuation of the fine conditions as we strolled along the beach with another hit list of plants in Nigel's back pocket. As the summer solstice had only passed the week before, there would be no rush to locate the plants before we lost the light either.

I remembered Newborough Warren well from my Bangor days. At the time I'd joined a practical conservation group tasked with carrying out everything from hedge laying and drystone walling, to scrub bashing and tree planting. And on one such occasion we'd spent an entire week constructing a set of steps to take visitors from the beach up onto Llanddwyn Island itself. Labelled an island as a result of being occasionally cut off by the highest spring tides, for the vast majority of time Llanddwyn was in fact little more than a peninsula. Approaching the 'island', I could instantly see that our feet would remain dry on this occasion too, and was utterly thrilled to see what appeared to be the very same set of steps I helped build still present a quarter of a century further down the line.

The first plant to come under our collective gaze once up on the island was one I recalled having seen here 25 years ago too, in the form of a large patch of maiden pink, which was just tucked behind a rock and a few metres from the main path. Due to the lateness of our visit, the flowers had unfortunately shut up shop for the day, with the result that on this occasion the maiden pink didn't appear quite as glamorous as I'd remembered it. But once down on my hands and knees I could still make out each petal's fringed leading edge and the distinctive greyish nature to the plant's foliage in what remained of the evening's light.

Logging a few more species along the way, and with Caroline present, our conversation slowly drifted from one of botany to that of our post-Covid hopes and aspirations for our respective families. And as the conversation flowed it was not until we noticed how much the light levels had dropped that we realised we'd lost track of time. Nigel had planned to end the day's proceedings with a couple more botanical treats in Newborough's large commercial forestry plantation, but with darkness almost upon us we'd have to be quick.

Planted after World War II, to both provide timber and to fix what at the time was seen as a problem with the shifting sand dunes, the large stands of Corsican pine are now as much a fixture at Newborough as those of the beach and dunes. Beloved by dog walkers and cyclists, the forest has become a source of great contention between conservationists keen to see this alien and largely sterile habitat vastly reduced in size and influence, and those intent on maintaining the status quo. A large-scale removal of conifers would restore a much-needed level of ecological balance, by helping to re-wet the dune slacks and certainly increase biodiversity across the whole site. But as the forest has also become enormously important for recreation, Newborough has found itself stuck in the middle between those fighting for nature and those demanding access.

It wouldn't take a genius to work out which side of the argument I favoured, but ironically one of the very few ecological upsides of the continued presence of the huge belt of alien conifers is that they also provide shade and cover for a couple of exceptionally rare orchids. Newborough Warren is one of just a few locations where dune helleborine can be encountered, with the other sites either confined to a few sandy locations in Merseyside, Lancashire and Cumbria, or at a number of scattered and contaminated spoil heaps across northern England and southern Scotland. The provenance of Newborough's few sword-leaved helleborines, however, is somewhat more dubious, with current thinking undecided as to whether their rhizomes were accidentally introduced with imported soil or have been here all the while.

Deciding, perhaps unsurprisingly, that she didn't much fancy a walk in a gloomy forest plantation at just after 10 p.m., Caroline opted to take the lighter and brighter route back to the car park via the beach, leaving Nigel and I to head into the trees. The forest at Newborough is believed to cover over 1,000 hectares, is criss-crossed by all manner of trails and was certainly not the kind of place I'd normally be wandering blithely into at dusk. But fortunately I'd not only brought with me an expert guide but also had a head-torch tucked in my rucksack, so took on the role of lighting the route while Nigel led the way.

With the forest now darker than felt comfortable, the thought fleetingly passed my mind that if I were suddenly left alone in the plantation then it was doubtful I'd even emerge before dawn. To put pastimes into perspective, botany doesn't have quite the same level of danger attached to it as, for example, hang-gliding or cave diving. But when tracking down some of our best and rarest plants, sometimes a little bit of calculated risk-taking was necessary, I said to myself, as memories of two close shaves with cliff faces and an incandescent security guard earlier in the year quickly came flooding back.

But as the botanical bloodhound alongside me suddenly stopped in his tracks to declare he'd found our first target, I was suddenly and gloriously reminded that risk-taking can also bring rewards. Dune helleborine usually flowers in the second half of July, which meant our visit was a touch earlier than ideal, but nevertheless the emerging floral spike on Nigel's find could clearly be seen in the glow of my head-torch, along with the orchid's typically yellowish-green leaves and diagnostic violet tinge to the base of its flower stalk. It was also a plant I'd previously managed to see in flower when living up here, when Tim and I had clambered around a fenced-off section behind Newborough car park's toilet block to find it back in the early 1990s. Botany can also, it would seem, take you to some dodgy places as well.

But we weren't even finished there, as in an entirely different section of forest Nigel then managed to locate the forest's second rare orchid. In contrast to the dune, the sword-leaved helleborine is at its best from mid May to early June, so all we could see instead of the snow-white blooms were a series of slowly maturing seed-pods. Having only seen this species a couple of times previously, I'd forgotten how tall and stately the plant can be, and was also able to admire the plant's long and decidedly er... sword-like leaves, which alternated up the orchid's slender stem. And the time we'd located the plant? 10.25 p.m. to be precise!

Looking at the date upon awaking the following morning, I suddenly realised it to be July 1st, which also symbolised the halfway mark in my big botanical year. This gave me six months to track down the final 380 plants, but with all the low-hanging fruit already picked, and the floral season racing by, success was still far from guaranteed. Thanks to Nigel's substantial help, I'd been counting on North Wales being a trip where I would score heavily, but with no more than two-thirds of a day left in his fine company there could be no let-up. So I was delighted and relieved at breakfast to find that despite our late finish the night before, Nigel was equally ready and raring to go.

The reserve Cors Goch, which translates from Welsh to 'Red Bog', was our last Anglesey stop before heading onto the mainland. Just a 10-minute drive from where Nigel and Caroline lived, the reserve was also of huge symbolic significance for its owner (and my ex-employer), the North Wales Wildlife Trust, which had originally been founded in 1964 with the express purpose of purchasing the site. The reserve is a complex mixture of lime-rich fen and acidic heath, which had formed as a direct consequence of ancient geological and then more recent glacial processes. Considered exceptionally rich for its flora, it was another of the reserves that I'd got to know reasonably well during my practical conservation days.

We hadn't even left the reserve's car park before Nigel was able to point out the day's first new plant, common gromwell, growing along the hedgebanks leading to the reserve's entrance. Stopping only briefly to appreciate its creamy-greenish flowers, we then headed straight for the base-rich fen where the reserve's more unusual plants would be located. And it wasn't long before the figure of 377 plants still to find began to drop even further, as the lime-loving great fen and slender sedge were added to the list.

Despite the subtle beauty offered by sedges, they were on this occasion quickly sidelined by the discovery of a far more charismatic plant: lesser bladderwort. Usually found in damp, uneven terrain such as ditches, the bladderworts – along with the sundews and butterworts – are another group of plants that manage to supplement their meagre diets by catching their own prey. Rootless and free-floating, the bladderwort would have been impossible to find had Nigel not spotted its tiny slender flower stem rising vertically out of the murky water. Examining the flowers is key for distinguishing a number of closely related species, and in this case the blooms' long, lower lip and short, stubby spur were more than sufficient proof to label them as 'lessers'.

The danger to any insect, of course, does not come from the plant's flowers, but from the minute bristles and small bladders on its submerged, thread-like leaves. Working like a vacuum, a trap door becomes instantly opened whenever an aquatic insect brushes against the bristles on the outside of a bladder, which then causes the creature to be pulled inside as the pressure equalises. So successful is this technique, that studies carried out on the closely related greater bladderwort have estimated that tiny crustaceans, larvae and worms trapped and subsequently ingested, may account for as much as 50 per cent of each plant's biomass.

But apart from the sedges and bladderwort, I'd already managed to log most of the other plants on offer at Cors Goch, so we decided to head instead for the Great Orme. Here Nigel had pulled out all the stops by arranging for us to meet the BSBI county recorder for Caernarfonshire, who'd kindly put aside some time to show us round.

We met Wendy McCarthy close to her home on the Orme. She appeared of a similar age to Nigel, but unlike both of us had never used her botanical knowledge in any professional capacity, having worked instead with social services in home care for 20 years. Entirely self-taught, and obviously living in a place with many great plants, when early retirement had suddenly presented her with the opportunity to devote herself full-time to botany, she'd clearly seized it with both hands. Nigel had told me there was no one alive who knew the plants of the Orme better than Wendy. And upon being introduced, she listened quietly and patiently while I provided a brief summary of my big botanical year before then pointing down at the pavement and informing me, 'You'll probably need that then.'

Looking down to where she was indicating, I could see a scrappy-looking plant in the gutter that superficially looked like common storksbill, but the leaves were all wrong. Sensing she was quickly testing my level of knowledge I stuck my neck out and uttered, '*Erodium*?' This on my part proved to be an astonishingly (and for me an unusually) smart move, as while not knowing exactly which species of storksbill it was, by using the scientific name of the genus, I'd managed to show that I wasn't totally clueless. Wendy congratulated me, before adding that the plant, whose full scientific name was *Erodium moschatum*, grew like a weed in the area.

Not wanting to ask what the plant's common name was, I waited until Nigel and Wendy had walked off ahead before surreptitiously looking in my BFF to reveal I was looking at musk storksbill. My guide described the plant as a scarce plant of 'bare or sparsely grassy places, mainly near the sea', and I had to agree that the plant's location appeared to fit the book's summary perfectly. But with so many other fine plants to track down, I could scarcely afford to waste my time with a coastal weed, albeit an unusual one, and so quickly re-holstered my guide before catching up with North Wales' finest, who were already lost in botanical conversation ahead of me.

Wendy's plan entailed taking us to some of the best limestone grassland the Great Orme had to offer, and along the way she took the opportunity to explain a little about the geology of the Orme and why it was so key to its fascinating flora. Lying just north-west of Llandudno, the huge headland is about 2km wide and sticks out almost 3km into the Irish Sea. Rising to 207m at the summit, the whole site is feted with designations: being an SAC, an SSSI, and a large portion of the Orme is now managed as a Country Park and Local Nature Reserve by Conwy Council. Composed of limestone, the rock was initially formed between 300 and 350 million years ago when most of North Wales lay beneath a shallow tropical sea. Weather, erosion and subsequent volcanic activity have all impacted the appearance of the rocks, which in due course became moulded by a series of much later glaciations to leave the landform that we were currently walking up.

The Orme has been occupied for thousands of years for the purposes of mining, quarrying, farming and as a World War II firing range. But what had inadvertently helped to keep such a large portion relatively intact had been the headland's many steep gradients and its thin, skeletal soils. The best sites are now well protected for their biodiversity, and as we walked up through the limestone grassland, I couldn't recall ever seeing common rockrose in such profusion – almost the entire hillside was painted yellow. Taking advantage of this vast nectar source were a few butterflies too, and a good look at one showed it in fact to be a female silver-studded blue. Probably the Orme's most famous butterfly, the population of silver-studded blues here is now thought by researchers to be distinct from those found on England's south coast, labelling them as subspecies *caernensis*.

With three pairs of eyes now looking, of which two belonged to proper hardcore botanists, the new plants started to come thick and fast. The bizarrely shaped moonwort was the first to be discovered, followed by two real rarities: spiked speedwell and white horehound. Suddenly I began to feel overloaded and was unsure how best to spend my time – should I focus on looking for plants, photographing them or just appreciating them? I decided that in the company of such terrific botanists, looking for new plants (which was normally half the fun) would on this occasion be a waste of time, so instead tried to focus my efforts on enjoying and snapping.

Wendy called me over as she pointed out the Great Orme's most famous plant, sprouting out of a small crag just below my feet. Wild cotoneaster is considered to be our only native cotoneaster, and even more dramatically the Great Orme is the only place in the world where it is known to exist. To hammer home its rarity even further, Wendy then went on to explain that prior to cultivating a further eleven plants from cuttings and seed, which were then successfully transplanted elsewhere on the Orme, the entire world population had been reduced to just six plants. Once common across the Orme, the plants had been deliberately dug up by collectors for their gardens in the nineteenth century, and systematic overgrazing of the Orme's plants by sheep, feral goats and rabbits had followed. Furthermore, I was looking at the very specimen upon which the description and name of the species had been based – otherwise known as the holotype.

While all the information was somewhat mind-blowing, I have to admit the plant itself was somewhat underwhelming, and I would certainly have struggled to separate it from a number of other alien cotoneasters also present on the Orme, were it not for Wendy's presence. It should be pointed out that midsummer is not exactly the best time to look at cotoneasters, as it is too late for the flowers and too early for the berries, and the apparent differences in leaf shape were – to say the least – subtle. Slightly bending my own self-imposed rule of only accepting species onto the list that I'd be able to subsequently and confidently identify myself, I was about to break another when Wendy asked me if I was counting *Hieraciums*.

The genus *Hieracium*, or hawkweeds, are a group of perennial herbs that look superficially quite like dandelions. Growing in a wide variety of places, but with a preference for infertile, rocky habitats, most hawkweeds seem to reproduce asexually, or by means of seed that are genetically identical to their mother plant, in a technique known as apomixis. This has resulted in the formation of clones, or populations consisting of genetically identical plants, with many botanists accepting these clones as species in their own right. Currently 262 micro-species of hawkweed have been accepted onto the British list, including the one Wendy was now pointing at, which was hilariously called confused hawkweed or *Hieracium britannicoides*, a micro-species considered endemic to North Wales.

A recent survey indicated there to be at least 390 *britannicoides* plants spread across eight known sites, and that was probably 387 more than the number of people actually able to identify it, I thought – one of whom I was with. For those keen to identify the plant, here's some help, and for the record I'm quoting from a 2021 paper in *British & Irish Botany* of which Wendy was one of the authors:

> *Hieracium britannicoides* is characterised by having rosettes with ovate, obtuse rosette leaves with subtruncate bases often with retrorse teeth and sparsely hairy or glabrous

above with few to numerous simple eglandular hairs beneath, the absence of stem leaves (or sometimes a very reduced ± linear leaf), the few-headed inflorescences with 2–6 capitula, the involucral bracts with few to numerous stellate hairs (especially near the base), numerous dark glandular hairs and numerous simple eglandular hairs, and yellow styles.

Hieracium britannicoides is intermediate morphologically between *H. britannicum* and *H. britanniciforme* and has at separate times been included in both of them); it clearly merits its common name 'Confused Hawkweed'.

Confused? …

My reasons for initially discounting different micro-species of certainly the dandelions, brambles and hawkweeds, was that they were simply too difficult for me to identify, and if I'm honest life is too short. But that didn't mean that I wouldn't be able to appreciate them, and especially given that our guide to the Orme also happened to be one of the very authors who'd actually helped describe the (micro)-species in the first place, then the very least I could do was honour her fine work by adding it to my list.

Heading back south, Wendy then took us across a playing field and through a small wood before eventually popping back out into the open and to what could only be described as a breathtaking view. From our location right on the south-west shoulder of the Orme, we could see Llandudno laid out below us to the south, while to the west the North Wales coastline and the Irish Sea stretched off into the distance, with the eastern tip of Anglesey just visible some seven or eight miles away. Also immediately below us lay a steep south-westerly facing bank, with the ample presence of limestone debris at the surface indicating that the soil here would be skeletal at best.

When chatting though all the possible plants that we might see on the Orme with Nigel beforehand, Nottingham catchfly had probably been the one I had most wanted to see. Having graduated from Nottingham University for my first degree, I was particularly fond of the city, and as my steel-erecting friend had said, any plant with a proper noun in its name has to be a good one. When originally planning my big botanical year, I'd been initially keen to go and see it in the county of Nottinghamshire, only to then (somewhat disappointingly) discover that despite having been picked as the county's flower, it hadn't been recorded there for years.

I told myself that all good things come to those that wait, as we slowly worked our way down the bank, spotting as we went. The steady stream of stellar plants continued here too, with dark red helleborine, ploughman's spikenard and hoary mustard all recorded before Nigel and Wendy called me back to a plant I'd just walked straight by without as much as a glance. Definitely past its best, on its few remaining flowers I could clearly see the deeply cut white petals rolled back on themselves like little springs and the incredibly long and prominent stamens so characteristic of Nottingham catchfly. Placing my nose next to its blooms, I also wanted to catch a hint of the strong floral scent, which is mostly produced at night to attract nocturnal pollinators, such as moths. But as the aroma trigger had not yet been released I was to be disappointed, leading me to briefly wish I'd been here both a couple of weeks earlier and a few hours later.

But being in such an amazing place, with phenomenal people and terrific plants, I couldn't remain disappointed for long. The only way it could have been improved would be with a beer and my family to enjoy the view. Furthermore, with me now needing just 345 species to reach the magical four-figure mark, North Wales and two of its inhabitants had helped me break the back of my challenge.

CHAPTER 12

Home from Home

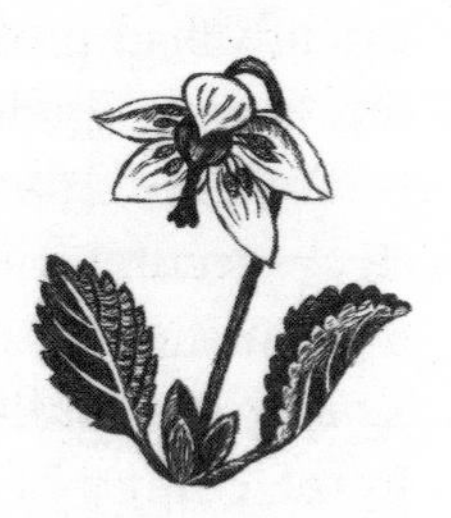

If all the different elements of my career were eggs in a basket, then close to half of them would now have 'Tour-leading in Scotland' stamped on them. This is somewhat different to four or five years ago, when the majority of eggs would have the immortal abbreviation of 'BBC' emblazoned across their shells. So as my career has developed, my guided trips to show guests the stellar wildlife of the Scottish Highlands has only taken on greater importance.

Courtesy of my 15 years on *The One Show* I must have visited the Scottish Highlands and Islands on at least a hundred separate occasions, to film everything from capercaillies and crested tits to basking sharks and black-throated divers. Moreover, I believe Speyside and its immediate surroundings represent the jewel in Britain's wildlife-watching crown. Habitat-wise this region encompasses extensive Caledonian pine forest, rolling moorland, rivers and lochs, the Moray coast and the mountains of Cairngorm National Park. My favourite lodgings also happen to be located right in the heart of Speyside, that being the Grant Arms Hotel, in the charming Highlands town of Grantown-on-Spey.

Usually the best times to come up to Speyside for wildlife are early spring and late autumn, as this is when the geese, ducks and swans are still here, the eagles are at their most active and you have a good chance of catching Speyside's fine range of mammals, such as pine marten, red deer and red squirrel. But aware that the fabulous Highland plants, butterflies and dragonflies also needed to be admired,

I'd also begun running a summer trip too, with twinflower and mountain ringlet as the stars.

Bidding a fond farewell to Nigel and Wendy, and with my profuse thanks for their time, expertise and patience, my next mission was to travel straight up to Speyside to lead one of the guided weeks both managed and run by the hotel. Due to Covid restrictions still in place up in Scotland, this meant that instead of guests being ferried to locations in minibuses they would instead be driving themselves to each location in their own cars, whereupon I'd then take up the reins.

While essentially working for the hotel, I also saw this as an opportunity to hugely expand my plant list with a range of arctic/alpine and northern plants simply absent south of the border. With the drive too far after my hard day's botanising on the Orme, I'd pre-booked an overnight stay close to both Carlisle and the M6, before then finally arriving at my home from home in Grantown the following day. A full itinerary had been planned by the hotel on my behalf, which included trips to all the forest, coast and moorland locations I'd become so familiar with over the last few years. But for me, the one standout day of the week would be when I took some of the hotel guests up to the Northern Corries (or Coires) in the shadow of Cairngorm itself. Here, a bucket-load of montane plants I'd still yet to see would hopefully be waiting for us.

The first few days were normal run-of-the-mill guiding, with relatively few opportunities to indulge in any serious plant-spotting, as the star-attractions of ospreys, bottlenose dolphins and golden eagles, had to be both found and shown to the guests. But with a clever bit of multi-tasking I still managed to winkle out a few excellent plants on these bird- and mammal-centric days, with the seaside umbellifer Scot's lovage and Caledonian forest specialist of intermediate wintergreen being the pick of the bunch.

Being so weather dependent, the Cairngorm visit was still far from certain to take place, so I was delighted when finding out at breakfast on 'mountain day' that the cloud base did indeed look high enough to permit a walk up to just short of 1,000m above sea level. Meeting up with all eight guests who'd declared an interest in some high-altitude botany at the main Cairngorm car park, the plan was to walk up to Coire an t-Sneachda, which is one of the most accessible of all Britain's high-mountain corries, to see what we could find.

Cairngorm Mountain's main car park has spent the last three years as little more than a building site, as the concrete stanchions supporting Cairngorm's funicular railway up to the Ptarmigan restaurant at 900m are replaced. Keen to head up and away from all the unsightly mess and noise, we shouldered our rucksacks before then heading south along the main path leading to the corries. With the path steadily heading upwards through montane moorland, it was fascinating to see the flora slowly and subtly changing as the altitude increased and conditions became more extreme. I'd walked this path a number of times and so had a pretty good idea where most of the best plants were, and as we all needed plenty of rest and water breaks this would also give us ample opportunity to both enjoy them and possibly find even more.

Stopping to point out a fabulous patch of flowering dwarf cornel to the group just by a path-side stream, the vegetation here was dominated by ling, bell heather and bilberry, with cross-leaved heath and sphagnum moss dominating any wetter areas. Now we were at an altitude of around 600m, the panoramic view of southern Speyside was laid out before us as I pointed out Loch Morlich and the combined forests of Glenmore, Rothiemurchus and Glenfeshie to the guests. Setting off again, the path then took us across the flank of the fell, while I kept my eye out for the point where we would branch off the main path for the left turn which would ultimately take us right into the heart of Coire an t-Sneachda.

I also knew this junction where the paths diverged to be a superb place to botanise, and leading the way gave me a brief opportunity, while the rest of the group caught up, to see what I could uncover. By the time everyone had joined me, I'd managed to line up a smorgasbord of botanical delights, with the undoubted star being lesser twayblade. Despite there being a few outposts in upland areas of England and Wales, Scotland can consider itself the true spiritual home of this delightful and diminutive orchid. Closely related to the much larger and more abundant common twayblade – which appears far happier in the home counties – the lesser twayblade, by contrast, prefers the splendid isolation offered by mountain moorlands, native pine woods and upland bogs. Perhaps best described as 'shy and retiring', lesser twayblades also tend to hide their light under a bushel – or in this case heather – which can sometimes make them difficult to spot.

The twayblades are so-named as both members possess a pair of oval or heart-shaped leaves that effectively form a basal rosette, but in the case of 'the lesser' these appeared to be positioned partly up the stem, almost as if the plant has produced them as an afterthought. The orchid's tiny little flowers sitting atop the stem also resembled minuscule elfin figures, while the whole plant (bar its paired leaves) was suffused with a delicate reddish wash. A female television executive I knew once hilariously described orchids as 'honorary mammals' due to their undoubted sex appeal. And judging by the 'oohs' and 'aahs' emitted by the guests as the pint-sized plants were pointed out, she was obviously onto something.

Ably supporting the lesser twayblade, bearberry, alpine lady's mantle, small cow-wheat and cloudberry were also present – with the last being the montane equivalent of blackberry. Unlike its lowland counterpart, however, cloudberry is always much shyer to flower and so only produces a minimal amount of fruit. Initially looking like red dewberries, they then turn a bright orange later in the season, and despite them having, somewhat unkindly, been called 'nowtberries' due to their indifferent taste, I've always found the ripest fruits to be capable of providing a pleasing sugary hit. But after a quick search the only ones we could find were still red, and as such, I suggested that perhaps they should be best left instead for any foragers passing by later in the month.

With everyone suitably refreshed after their break and impromptu botany lesson, we then carried on upwards. Almost with every vertical metre gained, from this point onwards, we could see heather's influence slowly waning, as the moorland habitat slowly gave way to that of subarctic tundra. As close to polar desert as we are ever likely to see in the United Kingdom, our immediate surroundings suddenly became dominated by bare scree, mosses and lichens. But with water at a premium in this environment, wherever it was present a flush of vegetation had seemingly risen to take advantage, and nowhere was this more obvious than at the point where the main stream emanating from the corrie above crossed our path. Looking both upwards and downwards along the stream, it was clear that the perennial presence of the life-giving water had effectively created a 'green seam', which was definitely worthy of further exploration.

A couple of years ago, when in this very same location, I'd manage to find a plant I'd only then succeeded in identifying several weeks later as alpine saw-wort, and so was keen to see if it could be relocated. While most of the group took five, a couple of the more-nimble members of our motley crew accompanied me as we slowly and methodically worked the stream's marshy margins. Just upstream of the path the most obvious flowers were provided by the golden globes of the aptly named globeflower, but in one small and particularly productive patch, where a tiny island of vegetation had formed in the middle of the stream, the twin beauties of starry saxifrage and alpine bistort were spotted clustered next to each other. Now on a roll, one of the guests then thrillingly found a couple of small white orchids, which I had singularly failed to locate elsewhere earlier in the week, before a couple of frog orchids were then picked up right along the stream. With heath-spotted and northern marsh orchids all seen closer to the car park, this brought the day's orchid tally to five, which had to be considered pretty impressive for such an extreme environment. Having seen a monster frog orchid only a couple of days previously on Nigel's roadside verge, I noted that the 'frogs' here, perhaps unsurprisingly, were shorter and squatter.

Something I'd come to appreciate when searching for plants was that the point when you're about to give up is precisely when they tend to turn up. Sensing that the group down on the path below were getting a touch cold, I was, likewise, at the point of abandoning the search when, hey presto, I noticed a small patch doing its best to hide from me. Like many other purple flowering plants, such as the knapweeds, scabiouses and heathers, which rarely bloom until mid to late summer, the alpine saw-wort obviously needed a couple more weeks too. However, in the most advanced buds a little hint of purple could be seen poking through, serving as a taster to the fabulous floral hit that the plant would soon produce.

'Cronk, cronk!' As we continued upwards to the corrie with the promise of lunch driving the group on to our final destination, a pair of ravens flew over, which astonishingly were only the second bird species, along with meadow pipits, we'd recorded during our entire walk. Cairngorm can of course be fabulous for birds, but to see the likes of snow bunting, ptarmigan and dotterel you frequently need to go even higher still to visit some of the mountain's more undisturbed areas, and be lucky. I'm also painfully aware that for many folk who come up with the express purpose of seeing Highland wildlife, birds may be their only interest, with even mammals (on occasion) not cutting the mustard either. But a quick glance round at the assembled group's tired but smiling faces told me that the flora and astonishing scenery had held their own, and our feathered friends (for once) had not been missed.

The long walk up to Sneachda is always worth it when you finally reach the lip of the corrie. From the top of the glacial cirque, the path quickly disappears into a huge boulder field, with a couple of small lochans marking the lowest point. Shaped like a bowl, the cirque effectively forms the apron of a stage, which is itself surrounded by a huge curtain of rock. At the headwall (or back) of the corrie, the sheer cliffs could also be seen rising up for what appeared to be a couple of hundred metres before eventually reaching the ridge line. Picking a huge glacial boulder from which to admire the view, we proceeded to pile into our packed lunches.

Aware that time will always be of a premium in such hard-to-reach places, I was incredibly keen to have an explore, but also needed to remind myself of my responsibilities towards the group, in an environment with which they were unfamiliar and where conditions can change quickly. In essence I couldn't just go off botanising, with a simple 'See you later!' A glance at the weather told me we'd be OK, certainly for the next half-hour, and after a quick verbal check as to the welfare of each guest, a debrief on the cirque then followed.

As the boulder field is such a remote and potentially hazardous place, even something as innocuous as a sprained ankle can quickly turn an excursion into an incident. From prior experience, I also know it to be surprisingly easy to become disorientated once in the boulder field, so I stressed that anyone venturing onwards with me should make it their responsibility to keep in visible contact with all members of the group at all times. With four keen to proceed no further, I'd also agreed that we would be back at the boulder with them in no more than 20 minutes.

Using the boulders as enormous stepping stones, I set out with the remaining four into the wilderness for one last botanical session. In such a harsh environment, the best places in which to look for arctic/alpine plants are often beneath boulders, as this is where the moisture collects, with the rocks offering additional protection from the biting winds. By using the technique of hopping across a few boulders before then stopping for a scan of any accessible nooks and crannies, this quickly came up trumps when a patch of roseroot was discovered.

Easily recognised by its rosette of blue-green fleshy leaves, each of the erect stems also formed a dense terminal cluster where the now-withered flowers were steadily turning into fruits. A member of the stonecrop family, the only connection roseroot has with its garden namesake is the strong smell of roses emanating from the roots when cut. One of the guests then managed to find marsh violet flowering underneath a boulder close by, while I eked out three-leaved rush, which also happened to be right alongside an even better plant, in the form of dwarf cudweed, which had made its stand in a small patch of bare scree.

The cudweeds, it has to be said, are not the most prepossessing of groups, and I've even heard them disparagingly called 'crudweeds'. But the group does have a mercurial charm, as most of the ten species (which are spread over two closely-related genera) are tough to find, often awkward to identify and only grow in really interesting places. Dwarf cudweed was a prime example, as on first impression it appears to be little more than a tufted, silvery and hairy blob. Only marginally more interesting when in bloom, its flowers are not much more than stubby little buttons. But the dwarf cudweed's interest comes not from its looks but from its fastidious demands. As a species it can only be found in Scotland, where it seems to favour well-drained, bare and stony mountain tops which retain the snow late into spring, before then becoming dried out in the summer – which all sounded suspiciously like the location we were standing in.

Deciding such a special plant warranted further inspection, I got down to the cudweed's level to appreciate its finer points with my eye lens, and was instantly glad I'd made the effort. Its narrow leaves were coated all over in a thick 'woolly jumper' of downy hair, to help no doubt with both the prevention of water loss and to give protection to the plant's more delicate parts from the endless frosts occurring in the higher reaches of the Cairngorm. It was a plant perfectly attuned to its environment.

One of the other plants I knew to be present in Coire an t-Sneachda, which also happened to be right at the top of my Highland hit list, was a decorative arctic/alpine called trailing azalea. Classified as a dwarf shrub, it is in reality an evergreen scrap of carpet that rises no further than 2cm above whichever rocky substrate it happens to be draped across. Possessing dense clusters of small, oval and evergreen leaves, it bursts into bloom between May and July when its rose-pink, starlike flowers stud the surface of the shrub. Painfully aware that this visit would probably represent my first and last opportunity to see the azalea all year, and having found it with comparative ease before, I lamented that on this occasion, however, it was nowhere to be seen.

With our 20 minutes almost up, I signalled to the group that we should now turn back to the others before they began to worry about us. Regaling the rest of the group, on our return, with the highs and lows of our brief foray amongst the boulders I said the only shame had been our complete failure in tracking down the azalea. One of the ladies who'd stayed behind asked what it looked like, and upon being shown a picture from my BFF, simply replied

'It's behind you!' in a voice that wouldn't have been out of place in a Christmas pantomime.

Almost standing on top of it, I don't quite know how I'd missed it, as the azalea had been no more than a couple of metres from where I'd taken lunch. In my defence, its presence was slightly less obvious without the flowers, which had already withered and been replaced by a set of little red capsule-like fruits... but only slightly less. Clearly a mature plant, it looked remarkably like a slow-growing bonsai tree that had just been steamrollered flat over a boulder. It also represented a super find to mark the highlight of a terrific day. All we needed now was to get everyone safely down off the mountain and back to the hotel in one piece, as only then would I be able to buy a round of drinks for the 'intrepid nine'.

The rest of the week passed uneventfully, as I concentrated on doing the job I was being paid to do, rather than progressing my own plant-based agenda. Bidding farewell to the guests after a fun-filled six days, by some careful wrangling I'd also managed to finagle one full day off to do with exactly as I saw fit, before heading home. And the only item on that to-do list was to see one-flowered wintergreen.

I can never quite put my finger on why I'm so much keener to see some plants more than others, but for no particular reason out of the 2,400-plus plants listed in the *New Atlas of the British and Irish Flora*, a select few do tend to leap out of the page at me, while others are merely glazed over. Certainly having never seen the plant before is a great starting point, as catching up with any plant or animal for the first time is always unforgettable. Next, if the plant has a certain level of rarity then that will certainly add to its cachet, especially if it grows in an interesting or challenging location. Finally, if the plant is exceptionally distinct or beautiful, that will help raise its appeal even higher.

So if I were to apply my entirely unscientific 'desirability coefficient' to a variety of different plants, then I can think of few that would rank as highly as one-flowered wintergreen. Traditionally a plant of old-growth pine forests, in June and July a small basal rosette of evergreen leaves puts up a slender stem no longer than a finger in height. From the single bud sat atop this stem, a white flower of impossible beauty then proceeds to unfurl. Confined to its own genus, making it a monotypic species, one-flowered wintergreen's scientific name of *Monesis uniflora* re-emphasises its incredibly distinctive looks. Here the plant's genus is a combination of the Greek *monos* meaning one, while *hesis* translates to delight. The specific name *uniflora* refers of course to its lone blossom, so just like 'New York, New York', this plant is so good it had to be named twice.

Also once known by the colloquial name of St Olaf's candlestick, small populations of this mythical rarity supposedly still exist in the pine forests of Speyside and the Moray coast. But despite being reasonably well connected in the Highlands' guiding world, all my efforts at finding someone who either knew where it was, or would be willing to divulge where to start looking had drawn a blank.

While the wintergreen appears to be struggling in the old-growth Caledonian pine forests, commercial pine woods have more recently become an unlikely saviour for this plant, with one plantation near Golspie in Sutherland believed to hold 90 per cent of the entire British population. I would normally have been reluctant to travel so far for any single plant, as I found it difficult to justify both the expense and carbon emissions that would be incurred. But on the other hand, having dove-tailed most of my plant-spotting excursions around other commitments then perhaps I'd just about earnt enough credit to allow for a one-off. As I wrestled with my conscience, what eventually sealed the deal was a quick look on Google Maps®, which told me the site was in fact much closer to Grantown than

I'd originally thought – Mohammed would have to go to the mountain.

A lady called Sue Williams coordinates all the wildlife-related business at the Grant Arms. Being not only responsible for running the celebrity guided weeks (one of which I'd just led), she also organises the rota for the wildlife guides working out of the hotel, and puts together a programme of evening lectures – all of which are a key draw for anyone staying there. Over the course of my numerous dealings with the hotel over the years she has become a firm friend, and during my many stays we will always try to fit in a walk at some point. These walks have the dual benefit of allowing us a catch-up while also giving her German shepherd Loki a good run out.

Deciding that my *Monesis* mission would be much more fun with a friend, I was delighted that Sue was able to accompany me, with the added benefit that she could use the visit as a recce for future trips from the hotel too. Much as I love guiding and sharing birds, mammals and plants with relative strangers, the job, like television presenting, can be all-encompassing, so in addition to spending the day looking purely for plants, it would also be a relief not to be on parade.

Leaving Grantown after breakfast, we headed across to the A9, before following it north towards Inverness. By-passing the Highlands' biggest conurbation, we then took the North Kessock Bridge onto the Black Isle before passing over in quick succession both Cromarty and then Dornoch Firth, with the latter firth marking the county boundary between that of Ross and Cromarty and the enormous county of Sutherland. By now I was in unknown territory, but it felt good to break free from the boundaries of the normal places I visited in Speyside and Moray. Before looking for the wintergreen, Sue suggested we give Loki a good run on the beach so that she would be much more relaxed by the time we got down to the serious business of flower-spotting. Loch Fleet nearby provided the perfect opportunity for canine exercise, and it being the summer we could let her run free along the sandy beach without fear of bothering any geese, ducks or waders.

When in the Scottish Highlands it is still blissfully easy to visit what is effectively a world-class beach without needing to share it with another soul. And while strolling along the golden sands, I was once again left to reflect that were we at a comparable coastal location in southern England, we'd probably spend most of our time dodging beachgoers, litter and dog mess instead. In fact the experience was so joyous that at one point it almost felt like the dog walk along the beach had been the sole reason for us coming here, but the plants would only stay on the backburner for so long. However, just before we were about to turn around, I spotted a small section of saltmarsh, which was too good to resist.

Checking the tide was still well out, I carefully stepped out into the somewhat softer sand and was almost instantly rewarded with what I thought might be a large patch of saltmarsh rush. Rushes, or *Juncus*, had been one of those genera I'd been terribly afraid of before the year had started, as they can all look very similar to the untrained eye. However, most rushes also tend to helpfully be confined to a particular habitat, and with the addition of both a hand lens and a little knowledge of which bit of the plant is called what, the prospect of identifying them had quickly turned from one of dread to that of keen anticipation.

In this case the inflorescence was terminal and each light brown capsule, or fruit, was around the same length as the darker perianth segments surrounding it. To confirm the equivalent of a botanical slam-dunk, the lowest bract when extended did not overtop the entire inflorescence, thereby eliminating the only potentially confusable species – which wouldn't have been able to survive in such a saline environment anyway… believe me, saltmarsh rush is much easier to identify than you might think.

Slightly closer to the water's edge, my first glasswort of the year was also present. Decidedly shiny and succulent, and looking like jointed pipe-cleaners, this edible plant is instantly recognisable to those who practise the fine art of food foraging. As a garnish on plates it tends to be confusingly called marsh samphire by chefs, and when steamed or boiled for a few minutes has a distinctively crisp and salty taste. The plant's salty succulence is of course an adaptation to enable it to survive in the salt-water environment in which it lives. Here the salts in solution prevent the plants from becoming sucked dry, or dehydrated, by the high osmotic pressure of the seawater. This concentration of salts is so high that the plants were even once used in the making of glass, hence the origin of their favoured name amongst botanists. So sought after is this marine vegetable nowadays that it often features on the menus of the smartest restaurants as a garnish with fish dishes.

With Loki exercised and happy, it was now time for the main event. Having been unable to find any more information about the precise location of the wintergreen in the forest meant that we'd have to use our super-sleuthing skills to track it down, which made me nervous. On the upside, by containing nine in ten of all wintergreens known, this was the plant's British headquarters, but on the downside the plantation did appear very large. If we failed to find it, it would have been a long way to have come for a dog walk, a rush and an edible plant.

Before plunging in, Sue and I took in a noticeboard, rather helpfully indicating a main footpath passing through the forest, which seemed as good a place to start looking as any. Surprisingly light for a plantation, the forest looked in much better condition than I'd initially feared, as a healthy understorey of bilberry, juniper and cowberry could be picked out, which all boded well, but it still looked like 'the one-flowered' wouldn't be one of those 'thirty-second finds'.

Sue, having just disappeared off the path to bag a number-two produced by her canine companion, suddenly shouted that she'd got it. Loki had only gone and found it for us. Joining Sue and Loki no more than seven or eight metres off the path, I instantly spied three or four plants with their heads coquettishly bowed, just like I'd seen in the numerous photos online. Sue and I agreed, in the understatement of the century, that they were absolutely beautiful, as I exulted in yet another plant that sailed into my rarefied 'exceeds all expectations' category. The flowers looked like little waxen lanterns and appeared in utterly pristine condition; we couldn't have timed our visit any better.

Falling flat into the downward part of a press-up position, I could immediately see the characteristically prominent and straight style, which was surrounded by a ring of ten curly stamens, laid out in five pairs. Each of these stamens terminated in an anther, an orange bed of pollen about the size and shape of a wheat grain. Placing my nose next to the bloom, I found that it didn't just look good, but smelled fine too, with a fragrance reminiscent of lily-of-the-valley, which appeared each spring at the bottom of our Chew Valley garden.

Down the stem, I could also see the pale, serrated and heavily-veined leaves of the basal rosette, which by the magic of photosynthesis would have helped power the production of both stem and flower earlier in the season. Like a number of our orchids, the wintergreen also benefits from a mycorrhizal association with a subterranean fungus, which helps the plant tap into all of the soil's available nutrients in return for a sugary fix.

Pollinated by bumblebees, the pollen of one-flowered wintergreen is extracted by the insects needing to beat their wings at precisely the right frequency to shake the pollen from its flowers. Otherwise known as sonication (or buzz-pollination) this astonishing technique enables the bees to collect pollen to take back to the nest as larval food, but in what must be a messy process, some will inevitably be transferred to the styles of other wintergreens during the gathering process. However, in what seems a monumental waste of effort, most of the seed appears to be infertile, leaving the plant little choice but to fall back on the tried-and-tested method of vegetative propagation by underground rhizomes.

A plant with such a high 'desirability coefficient' obviously needed to be photographed for posterity. So lining up my shot on a particularly attractive specimen, I was just about to press the shutter when the plant was monstered by a large doggy paw, as Loki instantly undid all her earlier excellent work. Our canine companion had obviously wanted to come and have a sniff too, and Sue was mortified as we looked at the pitiful remains of the stomped-on flower. But we shouldn't have worried; we subsequently found many more wintergreens close by. And it wouldn't be the end of the plant so cruelly, and accidentally, struck down in its prime – with its rhizomes waiting in the wings it could always resort to plan B.

CHAPTER 13

A Calcareous Road Trip

High summer often represents that period when the floral season quietly moves past its peak, before then beginning the slow, inexorable slide towards autumn. While acutely aware that time and plants would wait for no man, a few days at home were nevertheless vital for reminding my family who I was and to take a break from eating, sleeping and breathing plants. The problem with ambitious projects is that they can quickly become all-consuming, and the ones making the most sacrifices are those closest to you.

I was well aware that during my frequent plant-related absences my wife Christina had been left to juggle her own gardening business, our household, our son and half a dozen assorted pets. So there was little doubt that without her hard work, good humour and generosity, I'd have probably seen only a tenth of the plants I'd currently managed to log so far. For weeks, plant-spotting had taken precedence above all else, but upon slotting back into the family routine, it felt good to turn the tables, albeit briefly, as I devoted my time to those that mattered.

During this period, the plant-spotting didn't stop completely, but instead became temporarily relegated to opportunistic moments. My daily dog walks around the Chew Valley, for example, were still useful for ensuring that no common or widespread plants were missed as and when they came into flower. The time at home also happened to coincide with my mum's birthday too, culminating in a whole day of festivities in Bristol. But by getting up a touch earlier than usual, Zachary and I were still able to manage a quick 30-minute scramble around in the Avon Gorge to catch up with the endemic round-headed leek – known locally as the Bristol onion. With the botanical monkey off my back I was then able to turn up to my mum's party with a singular focus on making her day as special as possible.

Even before I started plant-chasing, going away has always been part and parcel of my job, and all too quickly a long-standing work commitment in Kent came around. The important difference here was that with my batteries now fully recharged and my family's blessing I felt ready to hold my nose and jump back in for the final push.

Early on in the planning process I'd realised that a key factor in accumulating my monumental total would be to cover as many different vegetation types as possible during the spring and summer months. Chalk downland and farmland, for example, were two essential habitats that I felt I'd still not yet properly covered, giving them the potential to deliver big returns. Remaining keen to ensure that my plant trips were kept as carbon conscious as possible, the opportunity to use my journey across to London to tack on a plant excursion also made perfect environmental sense.

The location I'd identified as most likely to substantially boost my list was a site called Ranscombe Farm, the flagship reserve of the wildflower charity Plantlife. The site is widely considered the best place in Britain to see a vast array of rare arable weeds, and still one of the most under-represented groups on my list. At such a large reserve, help on the ground would be vital, and so I'd arranged to spend a few hours with Richard Moyse, the recently retired warden of the reserve. I'd only visited Ranscombe once before, when asked to help celebrate National Meadows Day in 2015, and Richard had been tasked with giving me a guided tour. On that occasion I'd been blown away by his passion for plants, and so would once again be able to count on having the pick of the botanical crop to help me on my mission.

By clever coincidence, a drive to Kent from the West Country also happens to take you right through the county believed to hold an astonishing 75 per cent of all the United Kingdom's remaining chalk downland: Wiltshire. Following on from my successful trip to see one-flowered wintergreen with my friend Sue, I was keen to carry on the 'two's company' theme by inviting along an old pal of mine. In his capacity as presenter of BBC Radio Wiltshire's breakfast show, Ben Prater would often invite me onto his programme when needing an (informed) opinion from a wildlife expert.

So a quick stop-off at the fine reserve of Calstone and Cherhill Downs with Ben then ensued, as he interviewed me about the importance of Wiltshire's chalk downland, which also allowed me to add a few more plants to my list. As someone who works in the media I'm always disappointed at how little publicity wildflowers ever seem to receive, when compared to that of cultivated garden plants. So it felt good to redress the balance a touch, as with Ben's help we educated his listeners on the delights of stemless thistle, squinancywort and round-headed rampion.

Having reminded the good folk of Wiltshire as to the botanical importance of their chalk downland, I bid farewell to Ben, before pointing the car towards Kent, and my weedy appointment the following morning. The main artery taking you from the West Country to anywhere in south-east England is of course the M4, and as such was a road I knew incredibly well. Cutting across country along the A4, I then picked up the A338 for the short distance north, before planning to link up with the motorway at Junction 14. Coming from a wholly unfamiliar direction meant I'd never joined the motorway from what is known as the Hungerford turn-off and while queuing to enter the roundabout cast my eyes down to the road verge. It was cloaked in wildflowers.

'The unofficial countryside' was a phrase coined by the great naturalist Richard Mabey in the early 1970s to describe the crumbling city docks, inner city canals, car parks, sewage works and gravel pits across Britain where he first noticed wildlife flourishing apparently against the odds. Certainly roadside verges must also be included in this motley crew of highly modified sites too, with research recently carried out by Plantlife suggesting that collectively these linear nature reserves are equivalent in area to all our remaining lowland species-rich grassland. They can be incredibly diverse too, with Plantlife's surveys finding close to 700 different species of wildflower along roadside verges, which was not far short of the grand total I'd now amassed across a number of our finest botanical sites.

During my extensive travels around Britain's road network, I was of course used to seeing huge swathes of ox-eye daisies, rosebay willowherb, cow parsley, nettles and hogweed, but just one glance down told me this lime-rich verge looked considerably richer than most and I simply had to have a piece of the action. Driving slowly round the roundabout and ignoring the M4 east turn-off I was supposed to have taken, I could see the grassland and scrub in the middle was worthy of an investigation too and so turned my attention instead to where I could park. After a couple more revolutions I spotted a slip-road just north of the roundabout, and – taking care to ensure I wouldn't get run over – launched into some roadside botany.

The plant that had initially caught my eye due to its garish, magenta-coloured flowers was an everlasting pea. Aggressively scrambling over anything on which it could gain purchase, this long-lived perennial was patently making its own bid for immortality. In a clear indication of how many naturalised plants have settled in the United Kingdom, of the four closely related species of everlasting pea recorded here, only one – the narrow-leaved everlasting pea – is believed to be native to the United Kingdom. The four different species can usually be distinguished by a combination of leaf shape, flower colour and whether the stem is winged or not, and after a quick consultation with my BFF I was happy to label this everlasting pea as 'broad-leaved'. The most widely distributed of all the everlasting peas, it was supposedly introduced into cultivation from southern Europe in the fifteenth century. First recorded in the wild in Britain a century later, it has since run amok on railway banks and waste grounds, especially around London. And by the looks of this plant, it appeared to be gearing up to take on the West Country too.

In addition to the pea, I also managed to identify goat's rue, wild parsnip, ribbed melilot, mugwort and hollyhock in between dodging the cars. Astonishingly the addition of these six new species, which I'd been able to add to my list in as many minutes, had only been one species fewer than I'd logged during a couple of hours scouring the pristine chalk downland of Calstone and Cherhill. I suppose the caveat needs to be added that out of the six I'd identified along the verge, only the wild parsnip was actually native to the United Kingdom, as opposed to all the seven on the chalk downland reserve, but the provenance of the flowers didn't seem to bother the pollinating insects – the verges were humming.

After a good night's sleep I was ready for Ranscombe ... So ready in fact that I'd arrived half an hour earlier than the time I'd planned to meet Richard. Situated just to the west of the Medway towns and a stone's throw from the M2, Ranscombe is one of Kent's best-kept secrets. The site comprises 260 hectares, which is fairly evenly split between ancient and semi-natural woodland and tenanted arable land that is now managed principally for its rare arable weeds. Now within the North Downs I was back on chalk again, and with the weather dial set to warm and sunny, I'd brought enough suntan lotion and water to last me until the middle of the following week.

Loitering around the entrance to the reserve, I killed time by taking in the floriferous banks running either side of the gravel track. Purples were in abundance here, with nettle-leaved bellflower, field scabious, marjoram and wild basil all quickly identified. But having already seen most of these plants elsewhere, wild basil was the only new species I was able to add to my list – which now stood at a grand total of 722. With cow parsley and rough chervil already over, I also managed to bag the third and last member of the roadside white umbellifers to flower, in the form of hedge parsley . . . 723.

Having initially turned up too early, I apologised to Richard for what had now turned into a late arrival at our pre-arranged rendezvous point, mumbling that I'd been distracted by plants at the entrance. A genial smile from Richard assured me I needn't have worried, before adding that plants can do that to a person. Looking like the archetypal plant warden, with a fine beard, a broad-brimmed hat and an eye lens dangling around his neck, Richard had also brought along his successor, a far younger chap than both of us, called Ben. Being relatively new to the role, Ben was keen to learn of the location of some of the reserve's more unusual plants, and following a brief chat about the plan for the morning we set off to see what we could find.

Walking first along a section of the North Downs Way, Richard told me that having been born and bred on the reserve's doorstep, he'd been visiting the site since he was five. He'd also delved into the site's history with old maps, and remarkably didn't think the field patterns had changed in 350 years. The path we were on marked the junction between a largely abandoned old hedgerow and an arable field, called Longhoes Field. Longhoes looked stuffed full of arable weeds, in a scene that would have been all too common in the seventeenth century, but has since disappeared from modern Britain.

Alongside the swathes of common poppies were other plants of a bygone era, and as Richard got down onto his knees he pointed out one of Ranscombe Farm's signature weeds: broad-leaved cudweed. Already familiar with the genus from previous trips to both the Brecks and the Highlands, my trusty eye lens revealed an attractive, silvery-green plant covered in white, woolly hairs with the top of each stem producing globular heads that looked like clusters of tiny pyramids covered in yellow bristles. As I tried in vain to find anything that appeared to be a flower, Richard informed me that they were there, but happened to be both petal-less and hidden from view. Having declined drastically in the last 60 years, broad-leaved cudweed has now been reduced to just eight sites across south-east England, with Ranscombe apparently holding more plants than all the other locations put together.

Richard suggested that there was something even rarer nearby, pointing to what appeared to be an unremarkable-looking brome-type grass with an array of awned spikelets attached in a somewhat haphazard fashion along the top of the stem. I was happy to admit that it looked like a brome, but other than that I had no idea. Richard revealed it to be an English endemic called interrupted brome, which had previously been declared extinct worldwide in 1972. Thought to have arisen in the nineteenth century as a novel species that had formed as a result of a substantial and abrupt genetic change, it then went into a serious and rapid decline as the replacement of horses by motorised farming equipment reduced the demand for sainfoin as a fodder crop. An improvement in seed cleaning methods then effectively caused a tail-spin from which the plant simply couldn't recover.

How it was brought back from the dead is a remarkable story, starting with a botanist called Philip M. Smith, who collected seeds from the last population of interrupted brome in Cambridge. Managing to successfully germinate the seeds, he then grew the plant in pots placed on his windowsill, totally unaware of its subsequent disappearance from the wild, before then presenting seed to several colleagues. Some seed derived from Philip's original stock also arrived at Kew Gardens' Millennium Seed Bank, where a project was initiated with Paignton Zoo to grow the plants in sufficient number, with the ultimate aim of attempting its reintroduction back into the wild. After only short-lived success with a reintroduction back into Cambridge in 2003, a more successful outcome was achieved at a site in the Chilterns, where the brome not only managed to geminate, but also fruited and persisted. This project marked the first successful reintroduction of a plant back into the wild in British history.

Richard then went on to explain that the specimens we were looking at were also progeny of the original plants first introduced into Ranscombe in 2015; proof that with a comprehensive understanding of the grass's ecology it can and does have the ability to stand on its own two feet. Talk about back from the dead! A masterclass on bromes with awns then ensued, as we compared the interrupted brome with rye brome and great brome nearby, none of which I was at all sure I'd ever seen before, up until that point.

Needing help with my brome identification was one of those sobering instances I'd already experienced on a couple other occasions, whereby despite having become proud of the huge gains I'd already made with plants since the start of the project, I still had some way to go before having the audacity to call myself a botanist.

As Richard and Ben strode out purposefully into the field while discussing the finer points of reserve management, I ploughed my furrow along the margin and was delighted to be rewarded with a plant in a very special group; one which was both rare and that I knew. I'd previously seen dwarf spurge before at a nature reserve dedicated to arable weeds in Somerset, and immediately recognised its grey-green colouration, crescentic horns and slender leaves. Shouting my discovery to Richard, I was brought back down to earth when he told me it was common here, but he'd just found one that wasn't.

As I waded through the herbage, gathering burrs as I went, Richard had a look of flushed excitement on his face as he pointed down to the smallest, most inconsequential seedling ever, declaring it to be pheasant's eye, to which I responded with incredulity that he'd found it at all. No more than 5cm in height, and sprouting directly up from a clear patch of chalky soil, the seedling's foliage *was* distinctly feathery, but as I would never have identified it on my own (or found it in the first place) I had to put it straight into the 'if you say so' category. So-called because of its scarlet-red flowers with dark poppy-like centres, the flower perhaps resembles an anemone more than a pheasant's eye – but that's names for you.

Always confined to the chalk and limestones of southern England, it was never a widespread plant, but common enough in the eighteenth century for it to be gathered for sale in Covent Garden Market under the moniker of 'Red Morocco'. Highly sensitive to modern herbicides, the only place this plant regularly appears is in disturbed ground that was formerly arable land on Salisbury Plain in Wiltshire. Elsewhere, the plant's allergy to twenty-first-century Britain continues, with the result that it rarely persists for any length of time, even apparently at Ranscombe.

Seeing impossibly rare plant after impossibly rare plant produced mixed emotions: on the one hand as a naturalist there is always a buzz seeing something regarded as gold-dust, but on the other I felt more than a touch of melancholy that in our relentless drive for higher food yields we'd somehow castrated the wild appeal of the countryside. In my humble and terribly biased opinion there needs to be more room put aside for plants like this – our farmed landscapes are richer for weeds.

As we steadily worked our way west, Richard announced we were now entering the richest part of the reserve, with some of the site's most famous plants, which surprised me, as I thought we were already in top gear. As if to illustrate this last statement he beckoned me over with what had quickly become my favourite phrase: 'Mike, come and have a look at this.' Some plants seem to know they're very special by possessing an aura, and meadow clary was clearly in this elite group. Instantly identifiable by its striking violet-blue flowers, which grew in distinct whorls up the flower spike, the flowers themselves were dead-nettle shaped, but larger, far more impressive, and with the upper lip shape strongly arched like a sickle. At the base of the spike were the root leaves, which were wrinkly, crinkle-cut and smelt of sage, but these received little more than a cursory glance and sniff as my eye was constantly drawn back to the bewitchingly beautiful flowers.

While picking my jaw off the floor, Richard explained that even though it is now only found in around 20 sites, it has always been considered a rare plant in Britain. Mostly found on the North Downs and across the Cotswolds, it appears to favour chalk or limestone soils in sunny open grasslands or occasionally along woodland margins, as appeared to be the case here. Apparently Ranscombe Farm was also the very first place that meadow clary was found when finally added to the British wild plant register in 1699 and is believed to have graced the site ever since.

It was a plant that I had seen in the Cotswolds a few years previously, but the flowers then had been slightly past their best. On this occasion, however, they were nothing short of immaculate. Needing to find a reason to stay with this gorgeous plant for just a while longer, I took out my camera to preserve its perfection for posterity, but as an act it only served to delay the inevitable. Believe me, meadow clary is a very difficult plant to turn your back on.

As someone obsessed by natural history I must also confess to an addiction, which is buying books on wildlife. By current estimates my book collection runs into a couple of thousand, and while some of these books have hardly been opened, others have been thumbed through to the point where they're falling to bits. *Britain's Rare Flowers* by Peter Marren comes into the latter category, and was one of the few reasons I still paid attention to flowers in what I now retrospectively called my 'birding years'. I always remembered a chapter Peter wrote in the book, entitled 'A Whiff of Ground Pine', where he records his travails looking for the plant he described as 'looking like a first-year seedling of Scot's pine, with a sharp whiff of resin to match'.

Ranscombe also happened to be one of the twenty-odd locations where ground-pine still existed, but this little 'desert' plant is never easy to track down, and on my previous visit to the reserve Richard had been unable to locate it. We'd now moved to the most westerly part of the reserve, called Kitchen Field, and with Richard now physically and metaphorically 'in the zone', hopefully today would be different. Beckoning me over for the umpteenth time to admire another plant, I hadn't realised *which* plant he wanted to show me until he mentioned that this was a case of 'second time lucky!'

Down on my hands and knees in an instant, it was joyous to finally come face to plant with a species that had first lodged in my mind – courtesy of Peter's book – over 20 years ago. Richard had located a clump of three plants, of which the tallest could not have been more than 8 or 9cm, and which were sprouting out of what appeared to be little more than the thinnest sliver of soil atop a small, open patch of chalky rubble. Intensely hairy all over, the plants' slender leaves looked like hairy pine needles and stuck straight out from the stem, somewhat like the bristles of a tiny toilet brush.

I'd also recalled that Peter had written that its flowers are produced 'rather grudgingly', and this was indeed the case here too, as only the largest of the three plants had one flower left in good condition, with a couple more looking well past their sell-by date. As I moved in for a closer look at the flower, my nose suddenly reminded me I'd forgotten to check the most distinctive feature of all: its smell. The pine smell is created by resinous oils, which lie within the plant tissues to dissuade various herbivores, like rabbits, from chopping the pine down in its prime. These oils apparently give the plant a bitter taste, but certainly not a bitter smell, as the familiar aroma of pine shower gel gently accosted my nostrils.

The flower, like those of all the labiates, and including the meadow clary I'd just seen, was bilaterally symmetrical. And upon closer inspection I was surprised to find tiny little blood-red spots in the flower's yellow throat – a feature I hadn't hitherto realised the plant possessed. As was the case with many of Ranscombe's other plants, the last 100 years have not been kind to ground-pine. The application of fertilisers and weedkillers, the decline in rabbits due to myxomatosis, and housing developments have all conspired to bring this plant and many of its ilk to the brink.

Bidding a fond farewell to the ground-pine, we then located a selection of plants nearby, such as blue pimpernel, rough mallow and Venus's looking-glass – all of which appeared to be disappearing down the same ecological plughole as that of the ground-pine. Once again that familiar feeling of being elated and depressed in equal measure washed over me.

Suddenly realising that I was also a touch dehydrated, I glanced at the time and couldn't believe that we'd been botanising for just over three hours, with barely a drinks break. It had not only been hot and dry but it had been intense too, and I'd learnt more about the precise needs of arable plants that morning than in my entire career as a naturalist. As Richard had an appointment in the afternoon, we slowly began looping back to where we'd originally met via a section of the reserve called The Valley. Richard had one last surprise for me, but the time we'd spent tracking down all the other plants meant we might not catch our last target species at its best.

Rough poppy is comfortably the rarest of all our archaeophytic poppies. Along with common, long-headed and prickly, rough poppy has been a fixture in the British farmed countryside for possibly as long as 5,000 years. Originally from the eastern Mediterranean, minute poppy seeds from all four species are thought to have been accidentally brought as seed pollutants amongst corn crops imported by Iron Age farmers. Being annuals, poppies are opportunistic by nature and where conditions suit they can proliferate quickly. This sees them flourishing after activities like ploughing or tilling, for example, and was also the reason the churned-up battlefield at Ypres turned red from poppies during World War I.

While common and long-headed poppies are still a familiar site in unsprayed arable fields, rough and prickly have fared far less well due to their higher susceptibility to increased levels of nitrogen and the use of herbicides. Having managed to find prickly poppy with my pal Nigel in Norfolk back in June, I really wanted to bag the other rarity too. However, a peculiarity of rough poppy would potentially also make it one of the most difficult plants of all to find, in that the blooms tend to unfurl their petals at first light before then shedding them by midday.

Taking us almost full-circle back to the car, Richard announced we'd arrived at the area where the poppy was most frequently recorded, and with limited time for searching we fanned out to cover as much terrain as possible. Having seen virtually everything that we – or should I say Richard – had looked for, I was desperate to make a contribution to the collection of discoveries. But with the field thick with weeds, it was hard enough to see the ground at all, let alone look for a little pile of crimson-red petals.

Five minutes, and then ten minutes of intense searching passed, and I could tell our chance was slipping away. By now

I was a good 10m off the path and up to my waist in weeds, and deciding to quit while we were ahead, slowly fought my way back out to join Richard and Ben, who had given up the ghost too. With countless seeds stuck to my boots and a few scratches to add to the sunburn on my neck where I'd obviously missed with the lotion, I felt these minor inconveniences were a small price to pay for the glorious plants we'd seen. Hot and tired by this stage, I stumbled towards them before realising I'd somehow managed to drop my varifocal glasses somewhere in the field. Unable to see much without my specs, I felt a sense of panic rise up quickly as looking for glasses in such a cluttered environment would be nigh on impossible – were I to retrace my steps, I'd be far likelier to tread on them before locating them.

Berating myself for not having been more careful with such an integral piece of equipment, I tried to calm myself down by reasoning that as I used them constantly, they simply had to be close by. Thanks to the parted vegetation, I could clearly see my route through the weeds, even without my glasses, and after carefully replacing my feet, started scanning the ground – more in hope than expectation.

Suddenly, my eye caught a tiny patch of out-of-focus scarlet littering the floor, and squatting down to investigate further I found my glasses – unbelievably ¬ right next to the dropped poppy petals. Incredibly, the lenses and arms had combined to form a 'V', which appeared to be pointing right at the petals like the head of an arrow. But ironically it had been the petals' startling colour that had led me to the glasses, not the other way around. With my specs so important to my mission and so damn expensive to replace, I almost cried tears of joy.

Recovering my composure, I gratefully took in the rough poppy's petals, which also happened to represent species number 751 – I was over three-quarters of the way there. Feeling ever so smug, I then took the greatest delight in calling out, 'Richard, come and have a look at this!'

CHAPTER 14

Help in Heaven

If my race to a thousand could be compared to that of a 4 × 100m athletics event, then with over three-quarters of the plants bagged, this point felt like the equivalent of having just been passed the baton for the final leg. All I theoretically had to do was to race for the line down the home straight. But to stretch the analogy a little further, instead of competitors breathing down my neck while I sprinted for gold, the gentle winding down of the floral season still had the power to derail my months of preparation.

It was perhaps a good job then that I was heading back up to the Scottish mountains, where summer arrives at higher latitudes and altitudes a little later. My trip up north had been carefully designed to fit around a second day's broadcasting for BBC Radio 4's *Costing the Earth* strand about how our changing climate could impact Britain's flora. The first day's broadcast back in June had seen me interviewing Simon Harrap in the hothouse of Breckland, but the centrepiece of the whole programme would be a trip up a remote mountain on Scotland's north-west coast called Cùl Mòr, located just north of Ullapool, in Wester Ross.

Britain's arctic/alpine plants can be considered the canaries in the coal mine when examining the future implications of climate change. Often surviving in the coldest and least hospitable of environments, this specialised suite of plants will have arguably the most to lose if the elevated temperatures wrought by climate change mean that lowland species begin to take over. With this in mind, a visit to an upland region was considered essential for the narrative of the programme. Having pitched the idea originally at the end of 2020, I could have chosen any one of a whole host of mountains to highlight the plants' plight, but Cùl Mòr rose to the top of the list for a couple of reasons. Firstly I'd never climbed it before, and secondly it was the best-known location for an astonishingly rare arctic/alpine plant called Norwegian mugwort.

I'd first heard about Norwegian mugwort when chatting to Simon Harrap in the summer of 2020. Simon himself had made a pilgrimage to track down the plant in 2016. He, in turn, had been inspired by a quote in *A Colour Guide to Rare Wild Flowers* by the late John Fisher, who wrote, 'It is worth a week's hard labour to see the smiling face of this rarity, gazing like a miniature sunflower across one of the handful of sandstone boulders over which it presides.' Talk about a quote to titillate your taste buds.

Closely related to mugwort, which is an abundant weed across lowland Britain, Norwegian mugwort is not only far prettier but also much rarer. Known only from three mountain tops in Scotland, in addition to disjunct populations in Norway and the Ural Mountains, the Scottish plants are considered sufficiently different from their continental counterparts to elevate them to an endemic subspecies: *scotica*. Astonishingly the plant was only discovered as late as 1950 by a birdwatcher up on Cùl Mòr, a mountain rising 849m above sea level. Just short of the requisite height to be considered a Munro (that of 914m, or 3,000ft), the mountain's isolation makes it nevertheless impressive and imposing in equal measure.

Needing a contributor to bring the radio broadcast alive I had dipped into my mental address book of contacts from North Wales, and contacted Dr Barbara Jones to see if she fancied joining me. Working as Countryside Council for Wales (CCW)'s upland ecologist, until her retirement in 2011, no one knew the upland plants of Snowdonia better than her. I first met Barbara as a master's student at Bangor, and aware that her doctoral thesis had been on the ecology of the Snowdon lily, had even interviewed her about the plant while both of us hung off a rope in Snowdonia for *The One Show* around a decade later. She was in essence great company, incredibly knowledgeable and as fit as a fiddle (despite being at least 10 years my senior), and I was ecstatic when she said she'd be only too delighted to talk about flowers on the radio.

Ever the glutton for punishment, I'd arranged for the trek up Cùl Mòr to be just the first of a trio of ascents, as I'd then travel straight down to join two back-to-back BSBI field trips visiting the fabled botanical locations of Glenshee and Ben Lawers in the Breadalbanes. It would be a busy week. Unfortunately, due to the amount of travelling between various locations, taking my car up was the only viable option, which made for an incredibly long journey before finally arriving at my guesthouse just north of the small port of Ullapool. Barbara and her husband Nigel had taken a few days to drive up in their camper van from North Wales, with Cùl Mòr just the first part of a Highland trip that would also see them bagging a few Munros and catching up with friends along the way.

As the weather can be notoriously fickle on Scotland's west coast, we'd planned to give ourselves a choice from two possible days in which to climb the mountain, but upon meeting them both for dinner in Ullapool, an advanced forecast suggested we might just be in luck for the following day. As climbing Cùl Mòr could never be described as a walk in the park, the weather would play a critical part in the success of the mission. Starting off from a lay-by along the A835, I had been warned that the path soon petered out to leave nothing more than a stalker's track. A little further up, even this then disappeared too, with the route to the summit marked solely by a series of cairns. Negotiating a huge boulder field comprising scree, crags and even cliffs would not normally have bothered me, but white-outs occurred frequently on Cùl Mòr, which didn't sit easily with my lifelong affliction – that of a poor sense of direction. Barbara was, however, hugely experienced and exceptionally comfortable in the mountains, so we agreed that while I concentrated on the radio recording, she would focus on where we were going – the perfect division of labour.

Meeting up with Barbara and Nigel the following morning at the lay-by, I was already more than a little concerned that the summit of Cùl Mòr, which had been clearly visible at sunset only the day before, was now blanketed with cloud. I'm never normally one to panic, but a lot of planning had gone into this one day, and the weather had the capacity to throw the most enormous spanner in the works. But as a seasoned hand, Barbara had seen it all before and suggested that we should have a cuppa in their camper van instead, while waiting a while to see how the weather played out.

Now with a hot beverage we pored over the various online weather forecasts, all of which seemed happily to be of accord that the weather would slowly improve. But this seemed to contradict what we could see outside the van's windows, which was a continuation of the grey drizzle we'd been met with upon our arrival. I was just at the point of suggesting that we should try again tomorrow when Barbara stated with conviction that she thought we should just go for it. Given her years of staring at mountains, she thought the summit might just be clear, and quickly deciding to place my faith in her confidence (despite my own misgivings), I acquiesced as we loaded up.

It would be a long walk. And with time factored in to both look for the mugwort and to allow for frequent recording stops, we told Nigel, who would stay below, that he should not expect us back until at least 7 p.m. This would give us around eight hours on the mountain – more than enough time to find a plant and talk about it on the radio, we hoped. On top of food, water, waterproofs and an extra layer, I also had my camera, plant book and recording equipment and couldn't quite believe how heavy my rucksack was as we headed through the kissing gate and onto the path.

The first part was relatively easy going as we walked across the moorland. Stopping occasionally to record some chat, I asked Barbara about her passion for upland plants, where we were and what we were looking for. The mountain was largely composed of Torridonian sandstone, which had been deposited around a billion years ago, and apart from the mugwort, the mountain was considered fairly species poor. Surrounding us on the lower slopes was the usual mixture of ling, bell heather and bilberry, deer grass and the occasional splash of prostrate juniper, with little or no opportunity to add anything new to the list.

The first obstacle to the summit of Cùl Mòr was the stony rounded hill of Meallan Dìomhain, and as we started to properly ascend the chat between us fell away while we exerted ourselves for the first time. On the positive side, it had finally stopped raining, but as we climbed the cloud approached ever closer until it swallowed us entirely. Stopping to take a breather, I had also wanted to use the break to talk to Barbara about how the vegetation was steadily turning ever more barren, but we kept it short as the lack of wind meant that the infamous Scottish midges were annihilating us. This stop, however, did give me a new 'year tick' as Barbara pointed out the tiny and prostrate dwarf willow, complete with its fluffy catkins just beside what remained of our path.

Beyond the summit of Meallan Dìomhain, the stone cairns began to take over on the saddle, which would help guide the way between the two peaks. The problem was that it was still so misty we could only really see one cairn at a time, which necessitated taking a bearing until we reached the cairn, before then picking out the next one to stumble towards. After a short period on the flat we were climbing again, only this time it was much steeper as we worked our way towards Cùl Mòr's summit. In contrast to the mountain goat – that was Barbara – ahead of me, I was really beginning to blow as the walk then turned into a scramble over and around huge boulders. I was by now pouring with sweat, and had to additionally take real care not to tip over backwards with the weight of my rucksack.

And then it happened … I suddenly popped out above the clouds.

Above me I could see the summit of Cùl Mòr set against the backdrop of a cloudless, blue sky, and directly in front of me a beaming Barbara sat on a rock with 'I told you so!' written all over her face. This moment had to be recorded for posterity, and as I thrust the microphone towards Barbara she explained that we were currently experiencing a 'temperature inversion', where in simple terms it becomes warmer higher up. The views were simply astonishing, and just below we could see the blanket of cloud we'd just come through. To make it even more magical, the summits of the surrounding mountains of Stac Pollaidh, Suilven, Quinag and Canisp were all poking up through the cloud too, and looked for all the world like black icebergs floating past on a white sea. Where the cloud was breaking up a touch further west I could even see a hint of open water that was the North Minch.

At the mountain's summit, there was a small open-topped shelter, which we promptly ducked behind to both take a break and plan our strategy for finding the mugwort. My previous research had indicated that the stony plateau immediately to the north-west and just down from the summit was where the plants would be. Here the terrain was relatively flat, and effectively formed a dog-leg with steep-sided shoulders on either side. Being open to the elements, the sandstone on the plateau was in a constant state of erosion, with the result that the environment in between the boulders was almost reminiscent of a beach. It was amongst this sandy, gritty and billion-year-old substrate that I'd been reliably informed the mugwort would be found.

We strode down the shoulder together, with me recording as we went. The vegetation was so sparse up here that I hoped this would make the mugwort stand out – and so it proved. Stopping me in my tracks, while I was in mid verbal stream, a chuckle of delight from Barbara told me everything I needed to know – that all the planning had been worth it. My feet were instantly rooted to the spot as I gazed down in the direction indicated by Barbara's finger, to see a Norwegian mugwort in perfect flowering condition. I knew there and then that no other plant out of the thousand that I ultimately hoped to see would provide me with this level of pleasure and relief.

As we sank to our knees in synchrony, Barbara did a fabulous job of describing the plant's appearance. Proffering its head no more than 7 or 8cm above the rocky and sandy terrain, the mugwort's basal rosette consisted of a bunch of deeply and pinnately cut leaves, that were so intensely downy that they almost appeared to glisten silver. Looping out of this foliage was an even hairier stem that terminated in a single nodding flower, pointing coquettishly downwards, almost like it was too shy to show its face. No more than a centimetre across, the flower was around the size and colour of a yellow Smartie® with tiny disk petals, which in turn were surrounded by a tightly adpressed calyx of brown-edged sepals.

The mugwort just looked 'rare' and I could only presume the sole reason it had remained undiscovered for so long was that as Cùl Mòr had been labelled as being of 'little interest' to the wildlife-watching community, it had been simply overlooked.

Using the mugwort as our opening gambit, I then asked Barbara about what possible impact climate change might have on plants such as this, and was ultimately delighted that she, for one, was so positive about the future of Britain's montane flora. While declaring global warming a source of worry for all upland ecologists, she explained that a changing climate made the management of these sites ever more paramount. Many of our most important upland botanical sites are currently too heavily grazed, and were this constant grazing pressure to be at least lessened, as a result of tighter environmental regulation, then many montane plants would be better able to counter the impacts of a changing climate.

Being able to thrive in such a hostile environment means plants like Norwegian mugwort are as hard as nails, but where they do tend to struggle is when competing with other plants. Warmer temperatures may not intrinsically be a problem for arctic/alpine specialists, but where it does cause an issue is by inadvertently giving a leg-up to those lowland plants that struggle with extreme environments, yet are far more comfortable in a competitive arena. So prioritising the biodiversity of our upland environments should in turn provide our montane plants with a more suitable landscape to which they can hopefully move away from the lowlanders.

Interview concluded, Barbara and I then celebrated the mugwort's discovery with a couple of pieces of cake that Zachary had baked for us, and which I'd managed to tuck into a corner of my rucksack where they wouldn't become too squashed. Suitably satiated, we then began exploring the bouldery shoulder to see what else we could find. By now conditions were wonderful, with both of us down to just our base layers as the sun warmed our skin and illuminated the grins on our faces. The mugwort was surprisingly common too, and we quickly found a number of plants, with most appearing to be in flower and a few specimens even possessing flower heads with two blooms. Other plant species were scarce, however, with the only other species of note being a few yellowy-green cushions of a montane plant called cyphel – which amounted to my third lifer of the day, following on from the willow and the mugwort.

Having met a lady in the lay-by, who like us had been weighing up the pros and cons of climbing Cùl Mòr before apparently deciding to come back another day, we hadn't seen a single person up on the mountain. It was just us, the flowers and a world-class view, and I couldn't help feeling slightly self-satisfied that, in my opinion, everyone was missing out. Botany certainly has the capacity to take you to exciting places.

With both the photos and a world-class experience in the bag it was time to head down. We would have lingered a while longer, but due to an absence of phone signal all day it was important that we rendezvoused with Nigel to assure him all was well – in fact all was very well. Unsurprisingly, with far lighter rucksacks and aided by gravity, we were soon back on the stalker's path. Quickening our pace with more assured footing, something then caught our eye while crossing the path in front of us. Barbara exclaimed that it was a red grouse, before I had to point out that having been so accustomed to seeing red grouse in North Wales, she'd temporarily forgotten that the Scottish Highlands also have another species of grouse – that of ptarmigan.

It appeared to be a family party, and we watched as the parents took their four youngsters right past us before trotting off across the moorland. I'd seen 'ptar' plenty of times before on and around Cairngorm, but perhaps never at such close quarters. Such a stellar sighting of a difficult and elusive bird would normally have blown my socks off and instantly taken the mantle of 'crowning moment'... however on this one occasion, it didn't even come close.

In addition to the success of our day up the mountain, the other huge plus from having struck gold at the first attempt meant that the back-up day was free to do with as I pleased. Deciding to use it wisely, I headed for the north coast where, inspired by a previous visit for *The One Show*, I dropped in to a site called Dunnet Links to catch up with Scottish primrose – albeit on this occasion slightly beyond its best. I also happened to be pals with a chap called Dave Jones, who was the warden at nearby RSPB Dunnet Head. Having worked at

RSPB Titchwell Marsh together as young fresh-faced volunteer wardens back in the 1980s, Dave had remained in conservation while I'd sold my soul to television – but we'd both stayed in touch despite our contrasting careers.

Tipping me off as to the location of oysterplant on a beach nearby, Dave then met up with me afterwards to show me the RSPB's latest acquisition up in the Highlands – a site called Broubster Leans. Strikingly diverse in nature, with wet grassland, pools and rushy pasture, the reserve also had a population of a rare grass called narrow small-reed, which although not quite as memorable as the mugwort had just as much weighting on my list.

Stopping off with friends in Grantown-on-Spey on the way down, I was also well aware that the following two days would play a large part in deciding whether or not I would reach my total. A day each botanising on Glenshee and Ben Lawers had been in the diary for months. Both days were field trips organised by the BSBI for members to experience both locations in expert company, and for me represented two of my last opportunities to score heavily.

Glenshee was a wonderful day in fine company, as we first examined the mosaic of heathland and grassland with both acid and basic flushes, before then making the ascent to Cairnwell SSSI, where small sugar limestone patches and the adjacent limestone grassland held a huge range of montane species. Feeling, at times overwhelmed by so many new plants, I managed to add twenty species to my year list, of which an astonishing fourteen were plants I'd never seen before, including a total of four new sedges.

Perhaps the most memorable of the lot were almost the first and last plants of the day. Norwegian moonwort was only described as new to Britain in 2017, having originally been mistaken for its common counterpart, and yet found to be quite abundant right behind the Glenshee Ski Centre. Apparently Norwegian moonwort's leaflets have more deeply incised pinnae margins, or, in a language most of us can better understand, 'frillier fronds'. Having only ever seen common moonwort on a couple of occasions before, I was in no position to attest to the frilliness of its fronds, but it was however a plant of great curiosity value.

The other was located higher up on a lime-flushed grassland saddle, both above the sugar limestone patches and yet below the summit of Cairnwell. Alpine milk-vetch is simply one of those plants you have to see before you die. Known only from four sites, at which there are believed to be no more than twenty discrete colonies, the species is under constant threat from either overgrazing by sheep and deer, or trampling by the feet of wayward walkers. A member of the pea family, its leaves consist of a number of small paired leaflets, with a single unpaired leaflet at the tip. The short flower stalk is topped with a cluster of delicate pale lilac flowers, tipped with purple. It's simply a 'take your breath away' plant, and made me want to enact this ridiculous and fanciful idea of spending the next few weeks guarding it from teeth and boots until it set seed.

The other fine element of the day had been the folk I'd kept company with. Having spent most of my trips out either botanising in splendid isolation, with my family, or perhaps just one or two others, it was great to meet plant enthusiasts from a variety of walks of life, but with one facet in common: a fascination with flora. As most of the participants slowly learnt of my mission to see a thousand, the phrase 'Have you seen this yet?' was to be repeated to me any number of times during the day as the group showed themselves incredibly keen to help me towards my total. The group included a chap called Mark Hows, whom I found myself gravitating to more than most. Perhaps a touch younger than me, Mark's level of knowledge was commensurate with mine and he seemed equally obsessed. A veteran of a few more of these field trips, he immediately endeared himself to me when admitting over lunch that he had initially been cowed by the level of expertise present during his first trips out. But he had soon found out that, irrespective of knowledge, all were welcome.

I also discovered that he would be doubling-up with the trip to Ben Lawers the following day, so upon our arrival in Aberfeldy later that evening we met up for a delightful beer and a curry to chat about plants and so much more.

As I met Mark, the other participants and Dan Watson (our leader for the day) in the National Trust for Scotland car park, the good news was that the run of fine weather I'd experienced all week up in Scotland looked set to continue. Ben Lawers was certainly the last of Britain's fabled botanical stations that I'd never visited, and in my own mind Glenshee, despite bringing me many new species, had merely served as an appetiser to the main Ben Lawers dish. Once again, I recognised a couple of faces from Glenshee, in addition to Mark's, as Dan talked us through how he hoped the day would pan out.

Perhaps 10 years my junior, Dan had obviously been cast from a similar mould to that of Barbara, in that he was super-slim, weather-beaten and as fit as a fiddle. A Tynesider by birth, the last thirty-odd years of living in Scotland had been long enough for him to pick up a distinctive hybrid twang. Working for the National Trust for Scotland as their Highland Ecologist, his job – as far as I could work out – involved looking for and conserving rare montane plants. The day was nominally labelled as a training course for identifying grasses, sedges, rushes and ferns, but the safe assumption was that anything would be game as we slowly worked towards Ben Lawers' summit, which, at 1,214m, also represented Scotland's 10th highest mountain.

Before setting off, Dan was also keen to have a brief group discussion about the etiquette of picking plants.

Before the start of the year I'd always assumed picking wild plants was frowned upon, with the adage 'Take the book to the plant' considered the done thing. So I was initially mildly surprised to see experienced botanists picking plants all the time to inspect any features with a hand lens, rather than kneeling in a positively uncomfortable position to look at the plants *in situ*. Certain plants however, such as orchids, would never be picked, while most sedges were considered fair game. So there seemed to exist, certainly amongst experienced botanists, an intuitive understanding of which plants were 'pickable', even though paradoxically they might not know which species they were looking at until properly examined.

Strictly speaking, legislation under the Wildlife and Countryside Act (1981) makes it illegal 'to uproot any wild plant without permission from the landowner or occupier' in Britain, with 'uproot' defined as 'to dig up or otherwise remove the plant from the land on which it is growing'. However, you should not pick any plant on a site designated for its conservation interest, and nowhere was this perhaps more pertinent than at Ben Lawers, which was designated as an NNR, SSSI and SAC. Bearing in mind, however, that billions of wild plants have been destroyed inadvertently through agricultural intensification and development. I'd always found this rule slightly farcical, so I was pleased to hear Dan say that we would be able to pick judiciously, certainly lower down, but higher up where the real rarities lay, we should always ask before picking.

Setting off from the car park, we would be taking the route most often used by Munro-baggers, with the path following the Edramucky Burn, before then climbing over Beinn Ghlas towards the summit of Ben Lawers itself. Wheedling my way to the front of the line and directly behind our leader, just like a teacher's pet, I was keen to find out from Dan why Ben Lawers was so revered.

Walking as we talked, Dan explained that the exceptional montane flora was primarily a result of the seamless interlinking of geology, geography and climate. The mica schist rocks that Ben Lawers is renowned for not only supply the plants with calcium, magnesium, sodium and potassium, but also break down into a rich clayish soil that is able to retain moisture. These rocks also break the surface at the right latitude and altitude for a range of plants that are able to cope with being blasted by chill winds and covered with snow late into the winter. The mountain was not just important for its calcareous rock outcrops and crags, but also its high-altitude base-rich flushes, lime-rich grasslands and late snow-bed vegetation. All these habitats combined to make its arctic/alpine flora, as a whole, unparalleled in Britain.

Roughly following the burn, we were walking through what appeared to be hill pasture when I noticed a single flowering umbellifer with feathery and thread-like leaves at knee-height, and without a second thought picked it for a closer look. 'Which umbellifer is this Dan?' I called out, before a quick look at Dan's face instantly told me it perhaps should have been one of the 'ask before you pick' plants.

Gathering the group around, Dan revealed its identity as spignel, an umbellifer whose leaves when crushed produced a strong, sweet and aromatic smell, which I was able to confirm as delightful. In Britain the plant's stronghold was in Scotland, where it reaches its northern limit. Further south, it was considered uncommon in north-west England, while very rare in Wales and as such was now classified as 'Near Threatened'. Dan then rounded off his brief, impromptu lecture on everything you needed to know about spignel in 60 seconds, with a diplomatically worded invitation to pass the plant around for a smell, but a caution not to pick any more.

In my defence I hadn't actually uprooted the spignel, merely picking instead part of the plant, but it was a timely reminder for all of us to be careful, as we slowly ascended towards the 'rarity zone'. Periodically stopping to identify any of the commoner grasses, sedges and rushes, we soon found ourselves just above 900m, with the summit of Beinn Ghlas immediately to our south. Crossing the path at this location was a burn that Dan identified as being base-rich and deserving of further investigation, which gave everyone a chance for a breather and an opportunity to poke around its marshy margins.

Down on his haunches in a flash, Dan slipped effortlessly into full 'sedge mode' as stiff and russet were admired in turn. Russet sedge in particular was a plant with a very limited distribution that my *Colour Identification Guide to the Grasses, Sedges, Rushes and Ferns of the British Isles and North-western Europe* book described as growing in the Highlands of Scotland, especially in sites rich in calcium such as alpine bogs, flushes and rock ledges between 750m and 960m ... talk about ticking all the boxes.

Being one of the monoecious sedges, in that both male and female reproductive organs were present on the same plant, the single male spikelet was clearly poking above the two or three chunky female spikelets, which themselves looked like miniature Scots pine cones. Just like skinning a cat it appears there are a number of ways to identify a sedge, and the technique I preferred with sedges was to look principally at the colour and shape of their glumes (or scales) and utricles (or fruits) on the female spikelets. In russet sedge, Dan was able to point out the dark, blackish-purple and scarcely ribbed nature of the female utricles, which, combined with its location, was enough to seal the deal – and for me it represented another lifer, one more to add to the list and one closer to a thousand.

At this stage and location, almost every plant I couldn't immediately name appeared to be a lifer, as I was also able to gain my first ever views of alpine willowherb and mountain scurvy-grass at this base-rich burn as well. Saddling up again, we were now well and truly on the path to the summit, with the mica schist crags clearly jutting out to the south-west. Following the masterclass on sedges, one on mouse-ears quickly followed a little further along the path, with the confusingly-named arctic and alpine mouse-ears present almost side by side. Both species appear almost entirely confined to limey mountain tops, with the alpine mouse-ear marginally the commoner of the two species; the key feature to distinguishing these high-altitude cousins is by examining the nature of their hairiness. While alpine was covered in woolly, white hairs, arctic had far shorter bristle-like hairs. Perhaps the names of downy mouse-ear and bog-brush mouse-ear would not only be more suitable, but also more easily remembered, I suggested.

As we carried on ever upwards, the valleys either side of the ridge-line we were walking along were by now cloaked in cloud, in a similar temperature inversion to that I'd just experienced on Cùl Mòr. On that occasion, however, the foggy conditions had then been on the back foot, whereas on Ben Lawers the cloud appeared to be attacking us from the flanks. Constantly creeping up the valley sides in an attempt to engulf us, it was then partly beaten back by the sun in a kind of titanic good-versus-evil battle.

Finally reaching the towering mica schist crags, we veered off to the right of the path, with the plan to pick our way along the line of crags' lowest reaches to see what we could find both on the cliff bases and amongst the boulders. Time would be incredibly precious here, as I was painfully aware that we wouldn't have long at this botanical holy grail before needing to start our descent. To make matters even more urgent, the fog appeared to be gaining the upper hand, as visibility changed from 5km to 50cm, seemingly at the flick of a switch. With the pressure on, it was not the first time during the year that I was suddenly unclear how best to spend my time: should I be looking, making notes, photographing plants or listening to Dan? In the end I unsurprisingly tried to do all of them, with the result that I probably did nothing well. As I flitted from one amazing plant to the next, almost every species we stumbled across seemed to have alpine in its title: alpine meadow grass, alpine pearlwort, alpine forget-me-not and the best of the lot – alpine gentian.

Confined to just this location and some remote crags at Caenlochan, alpine gentian is a mega-rarity by any measure. It is also stunning, and as I gazed down with moistened eyes at the single (beautifully) flowering plant, sprouting at a height of no more than 6cm from a tiny, moist crevice, I also felt a measure of payback for the intense disappointment I'd experienced with its cousin at Upper Teesdale. Having missed the main flowering period of spring gentian, I'd only caught a glimpse of that bluest of blues, but the timing of our visit here appeared spot-on with 'the alpine' looking utterly pristine.

As the fog enveloped us once more, I tried to take a few quick photographs of the gentian through the murk, but could ill afford to linger as Dan called us over for a couple of high-altitude saxifrages, in the form of alpine and drooping saxifrage. While 'the alpine' was flowering beautifully on a moist rock-ledge, the drooping saxifrage had already gone over – but fortunately its distinctive tiny red bulbils dotted along the stem were very much present. These minuscule, vegetative clones of the parent plant also reminded me of the bulbils adorning the stems of the coralroot I'd eventually found after my reserve confusion in the Chilterns, and it was interesting to see the same propagation strategy adopted in habitats that were effectively polar opposites.

Keen to get us safely down off the mountain, and with the weather now on a knife-edge, Dan declared there would be just enough time to reach the summit before heading back down as long as we didn't keep stopping! Following his lead round back towards the ridge, Dan casually pointed towards a rocky slope to our left as being the best place to look for alpine fleabane. Too busy corralling all the participants he wouldn't have time to search, but I hadn't come all this way to turn down such an opportunity, and telling him I'd be 'five minutes', I shimmied up onto a reasonably accessible ledge and quickly out of sight.

Confusingly named, alpine fleabane is not even present in the Alps, meaning some textbooks call it by its other name of boreal fleabane. Almost entirely confined to south-facing mica schist in a few Scottish Highland sites, this montane rarity is very susceptible to grazing, meaning any surviving sites tend to be out of the reach of hungry herbivores.

Now easily 20m above the rest of the group, I quickly found myself on a ledge that was initially a metre wide before gradually receding to sheer rock face. And suddenly there it was ... clear as day and looking like a hairy, pink, supercharged daisy! Having found it myself it was immediately elevated, in my own mind, to 'plant of the day' and admittedly against some pretty tough competition. Hell, it was better than that – alpine (or boreal) fleabane had suddenly and out of nowhere crash-landed into my top ten plants of the year, and was even, conceivably, in the running for a podium finish.

Sticking my head over, I shouted down to see if anyone else wanted to clamber up too and was perhaps not surprised when my botanical buddy Mark was the only one brave and stupid enough to follow me up. He eventually made his way along the parapet to view the fleabane and was instantly delighted he'd made the effort. As it was a lifer for Mark too, we both got down to the serious business of carefully photographing it.

Confined to arctic Russia, northern Scandinavia, Greenland, Labrador and Newfoundland, how it had defiantly hung on here too at such a southerly location, and against all the odds, was anyone's guess. Mark and I for two, were just grateful it had – a brilliant plant to mark the end of one of my best day's botanising ever.

Dan's urgent shout from below brought us back to earth with a metaphorical – but fortunately not a physical – bang.

CHAPTER 15

We're Going on a Plant Hunt

I'd been away a lot and it was time to put my family first... well, equal first.

My extended trip to Scotland, plus a filming trip to Norfolk – via a quick detour to the Dorset coast to spot flowers with my friend Miles – had all fallen during my son's summer holiday. Left managing Zachary, her job and our home during my absence also meant Christina desperately needed a break, and so my suggestion of a father-son camping trip upon my return was met with a chorus of approval.

Taking Zachary away for a boy's only adventure appeared to please everyone. It would give my wife a chance to step off the incessant merry-go-round for a couple of days, while allowing father and son to spend some quality time together. It would also give me the opportunity to combine childcare and plant-hunting seamlessly, with the added incentive for Zachary of a change of scene – of which the highlight would be a night under the stars.

I'd been aware from day one that the key to reaching a thousand species of plant would be to cover as many habitats as possible during the critical period between May and August. And one habitat I still felt I'd neglected a touch was that of heathland. Obviously our family holiday down to the Lizard at the end of May had enabled me to track down many of the classic and widely encountered heathland species, but at this stage of the mission a spot of precise targeting was required. Heathlands also tend to be warm, sandy, look fabulous in high summer and as such exude the holiday vibe that I knew Zachary would appreciate too.

As far as heathland goes in the United Kingdom, the Isle of Purbeck down in Dorset is pretty unbeatable. Technically a peninsula, the name 'Isle' is thought to have originated from the barren heaths located between Wareham to the north and the coast to the south, which 'cut off' the land eastward so effectively it must have felt like an island. Purbeck is in fact bordered on two sides by the English Channel, while its north-eastern perimeter is delineated by Poole Harbour and the marshes of the River Frome. In addition to the golden beaches of Studland and iconic sites such as Corfe Castle, Purbeck also contains the largest area of lowland heath managed as a single nature reserve in England.

Historically, as a habitat, heathland has suffered from disappearance on a huge scale, with an estimated 80 per cent loss having occurred through land-use changes since the 1800s alone. What little remained had also became heavily fragmented, until some joined-up thinking by a number of conservation organisations on Purbeck came up with a bold plan to reverse this depressing trend. So in February 2020 Purbeck Heaths National Nature Reserve (NNR) was formed, which brought together the reserves of Hartland Moor, Stoborough Heath and Studland & Godlingston Heath to form a new 'super reserve', and it was here that I wanted to bring Zachary.

Top of my botanical hit list for the visit were the plants Dorset heath and marsh gentian, which hopefully wouldn't be too difficult to track down, as the last thing I wanted was for the mini-break to consist of little more than Zachary watching on while his dad botanised. I was, however, keen for him to re-engage with the project, which recently had become more of a 'dadventure' than a family affair. With Zachary and Christina on board, the only issue left to sort would be that of finding a campsite, which was easily arranged through the miracle of the internet.

❀

By the time a tent, sleeping bags, pillows, roll-mats, cooking paraphernalia, toiletries and a change of clothes had been packed, there was barely enough room to slot Zachary and I into the car for the two-hour drive to my chosen location of Knoll Farm Campsite. Situated slap bang in between Hartland Moor and Stoborough Heath, the campsite could not have been better placed to discover the delights of the peninsula's finest heathland.

Zachary was most excited about the camping aspect and upon our arrival was desperate to investigate the site and put up the tent before exploring further afield. The site was definitely more Glastonbury than 'glamping' and consisted of a large field – including all the usual bathing and showering facilities – along with a play barn for kids that contained go-karts, a huge wooden board game and a table-tennis table. Upon arrival we were welcomed by the site managers, a lovely couple who were also both in possession of the most astonishing set of dreadlocks. This then precipitated a hundred and one questions from Zachary about how dreadlocks formed in the first place, along with how best to manage their ongoing care and maintenance.

As the field was domed in shape, many of the best pitches around the edge were already taken, but eventually we managed to find a spot to erect our two-person tent close to a mature hedgerow that marked the perimeter separating campsite and heathland. I could not believe my luck, as fabulous plants would hopefully be little more than a wriggle out of my sleeping bag away. The weather will always make or break a camping trip, and with a little rain in the air I eventually acquiesced to Zachary's incessant pleading to go and play table tennis as we waited for the conditions to improve.

Fifteen games later and the weather had ameliorated sufficiently for father and son to take a late-afternoon stroll across the road and onto Hartland Moor. Stepping onto the reserve was as simple as stepping off the road, and immediately we could see the heathland's colours lit up beautifully in the late-afternoon light. Having spent a lot of time in the company of cameramen and women over the last 20 years, I knew their favourite time for filming to be during the so-called 'golden hour', which occurs just after sunrise and before sunset. Here the sun's low angle creates a softer, more diffuse light, with the effect that (in this case) the rich purple of the bell heather present almost seemed to glow.

As European gorse has largely peaked before the beginning of July, this also helped us pick out dwarf gorse, which generally delays blooming until high summer. As its name suggests, this rarer denizen of southern heathlands has a somewhat shorter stature than that of its ubiquitous European counterpart, meaning we had to be careful it didn't prickle our knees as we ambled along an old drover's track parallel to the road. Of course all the best heathland is not just a monoculture of heather but also encompasses a complex patchwork of habitats, and so it was good to see open heath, acid grassland, scrub and bog all juxtaposed here too.

Further away from the road, pools of standing water were also noted, with many surrounded by the tell-tale oranges and reds emanating from the moisture-loving duo of bog asphodel and round-leaved sundew, respectively. Deciding to see how far we might be able to get off-piste before the threat of wet feet drove us back, the first sign of an increase in moisture levels came from the sudden appearance of cross-leaved heath. Much better able to cope with its roots inundated than both bell heather and ling, this common fixture of wet heaths and moors is immediately identifiable thanks to its needle-like leaves forming up the stem in whorls of four below a cluster of rose-pink and urn-shaped flowers. Zachary, on being shown the difference between the three common heathers of cross-leaved, bell and ling, then said, 'well what's this one then?' as he casually pointed out the fourth ericaceous plant of the afternoon's work, which was also the rarest by quite some margin.

'Dorset!' I shouted. 'Well done son!' With his discovery of Dorset heath, Zachary had found one of my two key trip targets. Certainly taller than both bell heather and cross-leaved heath, and with a leaf structure more akin to that of a pine seedling, Dorset heath's fairly subtle differences appeared to have passed Zachary by. What he had instead noticed had been the colour of Dorset's flowers, which was of a subtly different hue to the other three: a touch darker than the pink of cross-leaved heath, yet a couple of shades lighter than the purple of bell heather. He had clearly inherited a keen sense of colour from his artistically-minded mother.

A few metres further on into the heath, I suddenly realised my tactical error had been not to put on our wellington boots in the first place, which instead of being on our feet were sat in the boot of the car. Ahead I could see that the wetter areas looked much more botanically diverse, and deciding they were too enticing to turn our back on, we ran back to the campsite to fetch our boots for a second stab.

As a child Zachary had always loved Michael Rosen's kids' classic *We're Going on a Bear Hunt*. And so, with our boots now on as we headed towards the boggy pools, I tweaked his fabulous prose to suit our purposes. 'We're going on a plant hunt, we're going to find a rare one, what a beautiful day, we're not scared!' By now the water, in places, was ankle deep, but I could see a great swathe of white-beaked sedge ahead of us, a sure-fire sign that the plants were becoming ever more interesting. Zachary, however, was by now beginning to get a little more fearful as to exactly where his dad was taking him, so I channelled both my inner Michael Rosen and his inner Bear Grylls by chanting, 'We can't go over it, we can't go under it, oh no! What are we going to do? We've got to go through it!'

Filled with this false bravado we splashed on, but it was not until I looked up that I realised we'd strode right into the middle of a floating sphagnum bog – with my botanical blinkers on once again I'd seriously underestimated the wetness of the situation. On the one hand the array of plants was seriously good, with bogbean, black bog-rush, Dorset heath and round-leaved sundew all around us, but the very instant we stopped to enjoy them was when we started sinking. Spotting the tiniest of islands, where a now-dead gorse bush had created a marginally drier hummock the size of a large serving platter, we were both relieved to step upwards and out of the bog to give us a breather and me a moment to work on our exit strategy. Zachary's boots were at least 10cm shorter in height than mine, and by the look of the watermark on the side of each of his boots he'd come perilously close to being overtopped.

'Dad, what are we going to do?' asked Zachary with barely disguised panic in his voice. At such a young age he'd become accustomed to the comfortable state of affairs where his parents were all-knowing and had an answer for everything, but on this occasion the scales had instantly fallen from his eyes upon seeing my indecision. 'Swim?!' I jokily suggested, to break the tension, which unfortunately had the opposite effect, as he was now on the verge of tears.

The issue was slightly exacerbated by the fact that the bog had already swallowed up our precise route in, and the extra weight from giving Zachary a piggy-back out would probably have sunk me too. Going further into the bog would be simply foolhardy, and the best option seemed to head for the nearest dry land, which in this case was a bank of gorse around 20m away and in roughly the same direction we'd come in from. Keeping a pace ahead of Zachary, my job – apart from berating myself for putting us in this position in the first place – suddenly became that of depth-tester, as I tried to direct him towards the marginally shallower sections.

As a technique for keeping my son's feet dry it worked admirably, but having taken one for the team, by the time we'd reached drier land, both my boots were already full of bog-water. Perhaps it was the least I deserved, I pondered, for allowing the 'green fever' to overtake my common sense. Compared to wading across the bog, navigating a path through the prickly gorse was a piece of cake, as we eventually stumbled back onto the track. Visibly relieved, Zachary's emotions then flipped from solace to glee upon seeing his old man pour the bog-water out of his boots before trudging back to the campsite.

'Pizza?' I said. 'Yes!' responded Zachary, in a manner suggesting that the unfortunate incident was already in the process of becoming erased from his memory banks. Changing into somewhat drier clothing, I'd discovered earlier that Vegan Pizza Fridays were a regular fixture at Knoll Farm, enabling us to – figuratively-speaking – fill our boots.

My first thought, upon waking the following morning, was to wonder why I'd put myself through another camping experience. My back was stiff, I was dehydrated, I'd barely slept and needed a pee. Zachary had seemingly spent the night in the arms of Morpheus, but his nocturnal disco-dancing moves meant that on the rare moments I managed to slip into a fitful doze, a sharp (and entirely accidental) dig in the ribs had succeeded in instantly rousing me.

Nevertheless, after a stove-cooked breakfast and a revitalising shower I finally felt ready to tackle the morning – which in this case entailed another marathon table-tennis session, which for some bizarre reason Zachary appeared to enjoy more than getting stuck in a bog. However, I eventually managed to coax the table-tennis bat out of his hand by promising a lunch on the beach at Studland in return for an hour's dry botany.

The western section of the Purbeck Heaths' new super-sized NNR belonged to Stoborough Heath, with its eastern perimeter on the opposite side of the road to boggy old Hartland. The plan I'd conceived, after poring over the map, was to walk up the road until reaching a track around 500 or 600m further up and off to the left, which would eventually bring us back via a circuitous route towards the campsite. The relatively flat topography and open nature of heathland always tends to create big skies, and with the wind having picked up overnight, it was hard to know where best to look, as both the scudding clouds above our heads and the flowers at our feet competed for our attention. Mercifully the path looked both well walked and much drier, and a handily placed information board happened to include a photograph of marsh gentian, to remind Zachary what we were looking for.

The south-westerly path initially took us though some acid grassland, which was of minimal interest, before reverting back to heathland proper, where purple slowly took over from sun-scorched yellow as the dominant colour form. Descending a touch, the path then passed a couple of boggy pools, which unlike the day before were more easily inspected from the dry safety of the adjacent path. Here, the aromatic smell of bog myrtle pervaded the air as the plants graded from tormentil and lesser stitchwort, which liked it hotter and drier, to the sedges and sundews, which preferred it warmer and wetter.

Not for the first time that year I tried to get into the mind of the plant. I had seen marsh gentian before in Anglesey's wet fens some 25 years previously, and seemed to recall a plant that preferred it damp rather than wet. Certainly the moisture gradient present between dry path and wet bog seemed to be covering all the bases. Gentian-spotting, however, was temporarily put on hold when Zachary spotted some movement out of the corner of his eye.

I always carry a transparent specimen pot with me when out on field trips, as you're never quite sure when it will come in handy, and after some fun and games Zachary eventually managed to pouch the hopping perpetrator. A quick check with my orthoptera app confirmed the red underside to the femurs and long wings as belonging to a large marsh grasshopper, the biggest and possibly most handsome of all our grasshoppers. It was also one of our rarest, with a range almost entirely restricted to the valley mires and wet heaths of the New Forest and Dorset – not terribly dissimilar to that of the Dorset heath Zachary had discovered the evening before.

Now on a roll, Zachary then went one invertebrate better, by spotting a raft spider sitting out on the water's surface of a boggy pool. It was large, chocolate-brown and with a pale yellow stripe running along either flank. Zachary couldn't quite believe it when informed this was one of only two spiders capable of swimming underwater, such as when chasing a tadpole, for example. What a shame I wasn't writing *One Thousand Interesting Invertebrates* instead, I thought, as this dynamic duo would definitely have been added to my list were this the case.

While watching the spider scrambling off through the vegetation we suddenly spotted the most delightful scrap of blue – before then declaring, 'That's it!' in almost perfect unison. Marsh gentian does not quite possess the level of rarity of its alpine and spring counterparts I'd seen earlier in the year. Its trumpet-shaped flowers are more subtle too, and by never appearing quite brave enough to fully open, it was definitely the shyest member of what must be considered an ostentatious genus. Nevertheless, the gentian-blue colour was unmistakable and despite the plant's attempt at being coy, deep down I suspected it knew it was nothing short of bloody gorgeous. 'What a lovely colour, Daddy,' said Zachary in a rhetorical swing of the hammer, which hit the nail absolutely flush on its head.

CHAPTER 16

Wading into Wetland Plants

I was flagging and my family were too.

When not carrying out the day job of talking and writing about wildlife, the rest of my waking hours were spent looking for plants, trying to identify them at the kitchen table or scheming as to where else I might be able to eke out a few more. My self-imposed target of a thousand had consumed me, to the extent that whenever outside, instead of walking head up and chest out, I had turned into this hunched figure with eyes glued to the gutters and verges.

A family day out to the Somerset Levels had taken me into the 900s. Here, the botanical pick of the bunch had been the frankly sumptuous golden dock, with a colour easily matching anything handed out at the Tokyo Olympics. But botanising in one of the less-visited parts of the Levels had also led to a couple of unsavoury moments. The first incident came when five large German shepherd dogs guarding a peat storage facility went rabid with rage while we walked past on a public footpath with our own dog. Barely held back by an inadequate fence, had the dogs managed to escape I couldn't bear to think what would have happened.

Still recovering from the near miss with the German shepherds, we were then told by a jobsworth at a private angling lake a little further on that we were trespassing, despite me pointing out to him we were clearly still on the public footpath. Ignoring our protestations he then tried to turn us around, which would have involved us running the gauntlet of the dogs again. Channelling my own inner guard dog, I snarled back at him to 'make me!' before, visibly cowed, he stood back to allow us through. Apparently the laws of trespass already exclude us from 92 per cent of the land in Britain, while some petty individuals still appeared keen to chip away at what remained.

Having scratched the heathland itch with Zachary, I now had to tackle the woeful under-representation of aquatic plants on my list. While trips to places like the Levels had undeniably helped, in reality I'd only picked off the easier and more obvious emergents, having conveniently labelled many of the plants below the water's surface as 'Too tricky'. But realising this was also where big gains could still be made, I'd booked myself on a wetland plant identification course.

Run by a lady called Sharon Pilkington, who at the time was BSBI recorder for both of Wiltshire's vice-counties, we met up at Langford Lakes in between Warminster and Salisbury for a crash course in aquatic botany. Under her expert tutelage the group learnt some of the tricks to help with identifying members of this hugely under-recorded group. Maybe the duckweeds, pondweeds and waterweeds weren't so scary after all. During the course of the day I'd also managed to boost my total by a princely 19 species, albeit with Sharon's help, and like any diligent student was also desperate to put into practice what I'd learnt.

For a while I'd been in email contact with a chap called Andy Byfield, a hugely well-known and respected figure in both botany and plant conservation. During his time at Plantlife, Andy had been involved in bringing an astonishing 'amphibious' plant called starfruit back from the brink of extinction. In fact starfruit's back-story was so remarkable that the plant had garnered my fascination years before the conception of my big botanical year.

So named because of its large, spiky and star-like fruits, which rise like Excalibur out of the water, the wild plant starfruit is no botanical relation to its dessert-forming namesake that became such a popular household dish in 1980s Britain. Starfruit was also at one time considered the most endangered wildflower in Britain. Though probably never common, this mysterious semi-aquatic annual could be found growing around the muddy margins of a number of small ponds across south-east England. Conditions that most appeared to favour the starfruit were at sites characterised by both fluctuating water levels and continual disturbance from grazing animals coming down to drink. Joining a long list of plants to have declined catastrophically since the end of the nineteenth century, it suffered greatly in the face of radical changes to grazing regimes and the drastic loss of ponds, none of which suited its antiquated needs. Tragedy was then compounded by ignorance, to such an extent that alarm bells with regards to its rarity only really began ringing in the 1980s, when hastily arranged surveys found it to be surviving in just two ponds, one in Surrey and the other in Buckinghamshire.

However, the plant's one saving grace was its seeds' apparent ability to lie dormant for decades while waiting for conditions to favour their return. This precise scenario occurred at an old site in Buckinghamshire, when a digger had been employed to clear away a mass of choking vegetation from a pond's perimeter. With the pond's bare banks now re-exposed, this inadvertent action had been just the opportunity that the starfruit's seeds, slumbering in the mud, had been waiting for, as the plant made a sudden and spectacular return to the site after an absence of 30 years. This ultimately led to the obvious and hilarious conclusion that perhaps the bulldozer was now the most valuable tool in the conservationists' armoury if this 'Critically Endangered' plant were to be conserved.

With the plant's precise ecological requirements now better understood, the conservation community set about replicating these conditions at a handful of starfruit's other old sites, while plants were additionally cultivated from seed before being translocated to a few 'ark sites' that appeared suitable. And while starfruit's future is still far from certain, this sterling work has not only given the plant a fighting chance of survival, but also presented me with a once-in-a-career opportunity to catch up with it myself.

I'd originally planned to go and see starfruit with Andy himself, but due to problems with his car, the trip would now be a solo mission. In a gesture of extreme generosity – and trust – Andy had however passed on to me the priceless information of three locations to check out: its original surviving site in Surrey and two other introduced locations elsewhere within the county. As luck would have it, a meeting in London gave me the perfect opportunity to fulfil two functions: that of business and botanical pleasure.

I have to admit to decidedly mixed feelings whenever visiting London, in that I'm always excited to go and yet quietly grateful to leave. So with my official business near Soho done and dusted, I headed towards Surrey's heathlands full of excitement and a measure of trepidation at the opportunity to catch up with a 'star in a reasonably small pond'.

Surrey's heathlands, much like those across Dorset and Hampshire, have become vastly reduced, with 85 per cent estimated to have been lost in the last 200 years. Being within striking distance of London, the remaining areas have become highly fragmented due to the extra pressure on land, meaning the best heathland was situated furthest from the capital, and in the west of the county. But I was heading instead towards a small scrap of heathland further east, which had become surrounded by the commuter towns of Dorking, Reigate and Leatherhead.

From the reserve's car park, a short walk took me through the heathland to the pond that had been starfruit's sole surviving locality in Surrey. Situated alongside a private track, the pond was no more than 20m across and 15m wide, and contained dominant stands of yellow flag iris and bullrush along both the track's side and the pond's eastern end. Elsewhere around the pond, clearance work recently carried out by the ranger service to create the more open and favourable conditions needed by the starfruit was clearly evident.

Unlike Dorset, where a miscalculation had left me with wet feet, this time I was better prepared with the correct boot-wear as I worked my way around to the more promising habitat at the back of the pond. This more open area, I figured, would not only allow me to inspect the muddy margin but also give the most unimpeded view of the open water. Empowered by my recent wetland plants course, I began to get to grips with initially what was sticking out of the water, with the incredibly distinctive branched bur-reed, lesser spearwort and water-plantain immediately visible. At this stage, no starfruit appeared obvious, so I busied myself instead with inspecting the plants along the shallow margin, where alternate water milfoil and water purslane were both quickly identified with my now enhanced skill set.

I was also pleased to find and successfully identify bulbous rush poking out of the water here too. This rushy species belonged to the group of plants that I would never normally have bothered to identify, as it was above my level of expertise. But on closer examination, the rush's bulbous stem-bases and tiny green plantlets sprouting from its inflorescences made it much more distinctive than I'd previously realised. As the *Juncus* genus does contain around 30 closely related species, it represents a group often ignored by beginners, but by tackling them head-on, and without fear, they were nowhere as difficult as I'd once thought.

While searching in and around the pond, I steadily came to the sorry conclusion that there were simply no starfruit plants to be seen anywhere. All the photos I'd seen and the literature I'd read indicated that the plant would either be growing in the drawdown muddy margin of the pond or sticking out of the water itself, but despite a second, much slower sweep of what I considered to be the prime area, I could only conclude that – certainly at this location – the starfruit had taken a year off. Maybe it had chosen to rest up in the muddy seed bank until the full impacts of all the restoration work had properly kicked in?

I was left with little choice but to enact plan B, which in this case was to trudge back to the car in order to visit one of the ark sites where starfruit had been introduced. Situated further south-west and much closer to the Sussex border, my chosen site out of the two Andy had furnished me with had been an abandoned quarry until its eventual conversion into a Local Nature Reserve (LNR). To enhance diversity, and as part of the Million Ponds Project – an initiative designed to reverse a century of loss and decline of this freshwater habitat – seven ponds had been created on the site in 2011.

With the ponds well established by 2013, a group of enthusiastic local botanists had then attempted an introduction of starfruit into two of the ponds using seed harvested from the plant's original Surrey site – where I'd just failed to find any evidence. And to everyone's delight the introduction appeared to work. Previous attempts at transplanting starfruit elsewhere could be described as 'mixed' at best, with short-term success invariably followed by a rapid decline in the following years. However, having been dug directly out of clay, the ponds' low nutrient status had obviously benefitted the starfruit by helping keep any competition in check. Further post-introductory assistance also came from volunteers, by ensuring the ponds didn't become dominated by bullrushes. This combination of suitable geology and proactive management had supposedly been so successful that Andy told me the plant had been described as 'locally frequent' in 2019. But with conservation work – in many cases – abandoned during Covid, would conditions still be optimal for such a needy plant? There was only one way to find out.

By now it was late afternoon and abandoning my car along the northern perimeter of the reserve I plunged in. The site seemed to be composed of mostly grassland and woodland, but with an eight-figure grid reference I was soon heading in the right direction. There was not a soul about as I took in the site, which I have to admit looked more like a prime dog-walking spot for locals than the holy grail for rare-plant hunters.

With the app on my phone indicating I was getting ever warmer in my grid-reference assisted game of 'hot and cold', a fenced-off area suddenly appeared. Its presence, I figured, must have been designed to protect visitors from accidentally blundering in and also potentially to give an incredibly rare plant some much-needed breathing space. Quickly scouting the fence's perimeter for 'keep out' signs, and following a sly look both left and right to check the coast was clear, I then shinned up and over. The vegetation inside was lush and quite tall, with the only indication that anyone had recently entered the exclosure being a trail of flattened grass, which led to a small pile of empty Budweiser cans next to the edge of one of the ponds.

To say the starfruit was 'obvious' would have been the understatement of the century. As I followed the beer-drinkers' trail to peer through the riparian vegetation, the plants' iconic six-rayed stars almost appeared to reveal their presence with a twinkle, and help brighten what was quickly turning into a decidedly gloomy evening. I uttered a strange gurgle of delight before then quickly counting what appeared to be nine separate plants in a pond no larger than 6m by 4m. My timing looked perfect too, as both fruits and the plant's distinctive three-petalled flowers were present on a number of the specimens. Unlike the earlier site, where I'd been searching the muddy margins, the plants here appeared to be emerging from the deep – making them decidedly more 'aquatic' than 'amphibious'.

The starfruit's narrow and yet oval surface leaves were splayed out like the hands of a clock around the plant, but due to the murkiness of the water it was nigh-on impossible to see their differently shaped aquatic leaves, which was the only feature not immediately visible. The plants were also of varying sizes, with the largest rising a good 30cm out of the water and containing at least 25 maturing fruits. I scouted around for the other pond, and this was soon located too, but despite being larger had only five visible plants. This took the grand total of plants to that of 14. The starfruit, certainly here, was in positively rude health.

As most of the plants were out in the middle of both ponds, I would have needed chest waders to get the best close-up photos, but the ones I managed to take appeared more than passable for someone who takes fine photos more by luck than good judgement. Pictures duly banked, I then took a moment to contemplate the plant's remarkable journey. Sitting cross-legged at the water's edge, staring at the plant's uniquely-shaped fruits, my mind also wandered towards those who had made the introduction happen.

We can be a very thoughtless species as we carelessly bulldoze habitats and species to obliteration, all in the name of developing Britain PLC. The State of Nature 2019 report painted a bleak recent picture, with 41 per cent of species in decline here since the 1970s and 15 per cent under threat from extinction. In the report, the United Kingdom was even rather shockingly called 'one of the most nature-depleted countries in the world'. However, the good news is that there are a number of dedicated folk who are patently not willing to just shrug their shoulders and accept the status quo, and in their own small way are actively trying to turn the tide. And certainly for their efforts here they had my admiration and gratitude because, rising out of the water like a phoenix, the starfruit's presence at this small suburban and unremarkable nature reserve had to be considered nothing short of a triumph.

CHAPTER 17

Going for Gold

Flushed with the success of seeing starfruit, I tried, through a process of steady accumulation during the second half of August and early September, to remain local wherever possible for the final push towards the thousand mark. Quantitatively, during this period, the most successful session entailed a day with the Wild Flower Society at one of their field trips to Grimley Gravel Pits in Worcestershire. Here, the pit margins and adjacent farmland helped bump my plant list up by an impressive 15 species.

Undoubted highlights of this trip up the M5 were a couple of different species of bur-marigold and both black and green nightshade, as I mopped up an assortment of plants in fine company. Amongst the motley crew assembled for the day's spotting were the BSBI recorders for the adjacent county of Warwickshire, John and Monika Walton, superb botanists who were able to help the leader Jackie with some of the more testing plants. Chatting about my challenge, they revealed they'd also previously attempted a 'big botanical year' themselves, but despite their best efforts and superior botanical skills had fallen well short of the thousand mark. 'I'm not there yet either,' I responded while John conducted a brief and impromptu masterclass on the identification of fennel pondweed.

Elsewhere a fern workshop close to home in the Mendips coordinated by Helena Crouch and Fred Rumsey – shining lights in the Somerset Rare Plants Group (SRPG) – helped bolster the list even further. Here a catch-up with Helena in particular was both well overdue and most welcome.

The Rare Plant Register, which Helena managed on behalf of the SRPG, was also crucial for helping me pick off an assortment of Somerset's late-flowering rarities. Here a solo day out saw me dropping into the Somerset Levels for marshmallow and arrowhead before then heading north-west for the slightly overrated slender hare's-ear and the hugely underrated sea wormwood. A subsequent family day out to Cheddar Gorge then added Somerset hair-grass to my list, which took me tantalisingly to 999.

For my thousandth shade of green I had three stipulations: firstly the plant had to be rare and gorgeous, secondly I had to see it with my family and thirdly it had to be found close to home. As I was scrolling through the Rare Plant Register one evening, a plant suddenly leapt off the screen that instantly fulfilled every criteria – goldilocks aster.

Restricted to just six coastal localities in western Britain, goldilocks aster is confined to limestone sea cliffs and rocky clifftop grassland overlying limestone, of which two of those sites were close to where the Mendip hills reach the Severn Estuary. Being a poor competitor and susceptible to heavy grazing, goldilocks aster has become a plant that has also been pushed to the edge, with inaccessible cliff ledges being the place (in many cases) where it has been able to hang on in there.

While the vertiginous nature of aster's preferred habitat gave me slight reason for worry, what eventually swung my decision towards attempting to make it my thousandth species was the fact that the plant is generally considered to be looking its best in September and October. When in flower, this limestone specialist has frothy golden-yellow flower heads that sit atop a slender grey-green stem, itself adorned with an array of super-slim leaves. Choice made, and family – keen to witness the epoch-making event for themselves – on board, we set off for the coast. We were on a mission to find some botanical treasure … we were going for gold.

Out of the two Somerset sites to hold a population of the plant, a small Local Nature Reserve called Uphill, just south of Weston-super-Mare, appeared to be our best bet. The other site, which supposedly held a far larger population, was the far more iconic Brean Down. But the goldilocks aster here was only observable from the northern end of Brean Down beach at low tide, necessitating the use of binoculars rather than an eye lens. I figured that trudging through silty mud would not go down well with the family and also the turning tide would add a time pressure that would only make the search more stressful.

Likening my race for a thousand to that of the Tour de France, I felt I'd already done the hard yards earlier in the year and so was keen for my final plant to be just like the last, or Champs-Élysées, stage. In the world's most famous cycle race it is customary for the lead on the final day to not be contested, which allows the riders crossing the Parisian cobbles to drink champagne while enjoying the sight and sound of their adoring fans cheering them on to the finish line. In sharp contrast, my audience would consist of nothing more than my wife, son and dog, but despite the lack of a welcoming committee I considered my achievement (were I to find the plant of course) no less than when Bradley Wiggins won the race in 2012, for example. And, much like Sir Bradley, it would also be something I'd never be attempting again.

This would be the first time any of us had visited the reserve, and as we passed through the entrance gates, the site seemed a curious mix of boatyard, marina and campsite, all set against the hugely impressive backdrop of a sheer wall of limestone cliff, with a church perched on its highest point. Obviously only the cliff and its immediate surroundings were of any interest to me, bar the car-park coffeeshop of course, and with batteries recharged we set off to track the plant down. As the aster was apparently restricted to only a couple of tiny sections of the reserve, I'd come equipped with some Ordnance Survey map coordinates and verbal descriptions of the locations, but with the sheer cliff face suddenly looming large in front of us, I had an instant sense of foreboding that finding the aster wouldn't be anywhere near as easy as I'd envisaged while sat comfortably in front of my computer back at home.

Coordinates can be incredibly effective on flat open grassland, for example, where you can access all areas, but on the challenging terrain of a cliff face, where large sections are simply inaccessible, then even getting close to where you want to be can be a trial in itself. The first place I was keen to look was inside an enclosure positioned at the base of the cliffs and alongside the Tidal Trail footpath, as 80 plants had apparently been counted there back in 2015. Unlike the enclosure protecting the starfruit, which had consisted of nothing more than posts and rails, this fence appeared somewhat more challenging.

As it appeared far too draconian to protect a rare plant, I wasn't initially sure if we'd be able to scale the almost 2m-high chainlink fencing, which was then topped off with a strand of barbed wire, even if we wanted to. But by working the perimeter I eventually spotted a weak point at the top of the gate, where the wire had been missed. The gate also had a 'keep out' sign, but having become inured to such notices during the course of the year, and which in my opinion were often of dubious legal standing anyway, I decided it should be treated with the contempt it deserved. With Christina not keen to attempt such a foolhardy move herself, and as we'd be unable to lift the dog over anyway, she suggested that she'd continue with Bramble to scout the trail further along.

By clambering up and sitting astride the top of the gate, I eventually managed to haul Zachary up alongside me, before then getting us safely down onto the other side. Mostly woodland and scrub, the enclosure inside was like a secret garden, with a screen of trees between us and the trail meaning no one could even see we were inside ... perfect! Following the merest hint of a path, we quickly reached the base of the cliff, which was lighter and brighter where the vegetation, unable to colonise bare wall, had come to a sudden and abrupt end. The GPS on my phone indicated the aster's position to be further along the cliff wall, and as the base of the cliff was impassable due to the positioning of a huge bramble patch, we scrambled up and onto a narrow shelf, which appeared to take us in the right direction.

Petering out into sheer rock face after no more than 10 or 15m, the shelf unfortunately appeared to be little more than a dead end, which left us with no option other than retreating from where we'd come. Not so easily defeated, Zachary and I then tried to see if we could get closer to the map coordinates – and presumably the plants – from around the other side, but this approach was thwarted by impenetrable vegetation here too. Last recorded here six years ago, I guessed that as the vegetation had shot up so much in the intervening period the asters might simply have disappeared upon becoming shaded out. Unable to get even close to where my GPS indicated the plants to be, we clambered back out of the enclosure to find Christina and to urgently rethink our strategy.

I felt that blithely following my phone's GPS, in this case, would lead us up more blind alleys, and we would be better served by looking for suitable habitat instead. Further along the path, Christina had discovered that the cliff gently sloped down to ground level, thereby making access to the vertical habitat more straightforward. By basing ourselves down there, we figured, we would at least have the opportunity to explore the rock face at a more accessible height, or potentially even climb up the slope to look for the goldilocks aster from above too.

With my notes indicating the aster had also been previously found at this more easily accessed end of the cliff face, it also occurred to me that for it to have been discovered in the first place then it must at least have been encountered at a location viewable from a place of relative safety. Down at this far end, the cliffs and cliff ledges were certainly easier to work, as I noted the yellows of yellow-wort and common ragwort, but any golden yellows were notable by their absence. Once again the map coordinates were difficult to interpret and the written descriptions were equally of little help as I struggled to ascertain which of the many ledges 'on a cliff ledge 10–20m up on a south-facing vertical cliff,' the notes appeared to be talking about.

Aware too that patience would soon begin wearing thin with the other members of the Chew Stoke Botanical Group, I suggested that perhaps we should try looking for the aster from above instead. Here a clear path meandering up to the church would take us back along the cliffs, but this time from above them rather than below. However, this option was not straightforward either, as looking for the aster now involved having to regularly peer over the edge, which sat very uncomfortably with my wife's dislike of heights.

About halfway up the slope a fence line, obviously placed to contain livestock or to delineate ownership, was present, bisecting the field before ending abruptly at the cliff. I lay down near a wobbly post at the end to scan for asters below for the umpteenth time. At this location the rock face appeared not quite as vertical as elsewhere. Briefly weighing up whether or not to squeeze round the post, I decided I didn't like the 3 or 4m drop waiting if I misplaced my feet, and also not wanting to unduly alarm my wife, I decided to look instead further up.

However with the phrase 'dogged determination' a shoo-in for the epitaph on my gravestone, I was still not ready to give up just yet. At Zachary's primary school, they hand out regular prizes to pupils for 'trying hard' and we'd developed a family joke about whether he would ever win the 'Perseverer of the Week' award. Upon taking one look at his father's steely determination, Zachary then obviously recalled this in-house gag before stating with a hilarity way beyond his years that 'never mind perseverer of the week, Dad deserves perseverer of the year!'

His timing could not have been better, as we walked up the hill howling with laughter at this perfect tension-breaker. Eventually reaching the church atop the cliff, I was now struggling to see where else to look for the aster, and without risking either mine or my family's lives I was stuck on 999 and out of ideas.

One of the many reasons I married Christina is for her way of looking at problems differently, and upon her suggestion that I should phone a friend, it suddenly struck me as the one and only blindingly obvious choice left. SRPG Helena would know! During the year I'd lost count of the number of times she'd helped me with regard to tracking down everything from purple gromwell to alpine pennycress and most of the goldilocks aster records on Somerset's Rare Plants Register happened to be hers.

Caught while working away in her garden, she sounded like she was glad for the break and told me that the plants in the enclosure were the hardest plants to locate and might conceivably not even be there anymore. She informed me that we needed to walk up along the clifftops and look for a wobbly post. 'Yes!' I exclaimed breathlessly, as I instantly knew exactly where she meant. Helena went on to explain that to find the aster you had to *carefully* work your way around the outside of the post, which she assured me wasn't quite as dangerous as it had initially appeared, and after skirting around the corner it then opened out to a ledge the size of an elongated kitchen table. And it was here that she'd seen the goldilocks aster in 2020.

Thanking my botanical saviour once again, I hung up as we all headed back down to the fence line for another go.

Shedding my rucksack at the base of the fence, I told my wife that if Helena had been able to navigate the post then she mustn't worry about me either. Having previously met Helena and her equally delightful husband Jim for tea and cake at their house, my wife knew her to be someone who was charming and sensible in equal measure, and so before she changed her mind I quickly shinned round the post and almost immediately dropped out of sight.

The ledge around the corner was warm and south-facing, and with so little other vegetation to compete with, the goldilocks aster was so immediately conspicuous that the act of spotting it almost caused me to topple over the ledge. By sheer bloody-mindedness and with the help of some inspirational people along the way, I'd only gone and done it. Avoiding the temptation of yelling 'Got it!' to my family waiting above, I decided instead to take a moment to drink in the plant's location. And what a view it had chosen for its (almost) last stand. From the lofty ledge, I could see the church up above, the Severn Estuary to the west and the Quantocks away in the distance and to the south. The goldilocks aster had chosen a millionaire's postcode.

Of the half-dozen plants on that precarious ledge, none were quite yet in full bloom, but despite their flowers being only partly open I could still clearly see a splash of golden yellow, and it took nothing away from the moment. A moment that needed to be shared with my family. Leaving the aster temporarily behind, I worked my way up, around the post and back into view, before taking in my family's imploring faces. Childishly and mischievously attempting to cut a dejected figure, I made them wait just a couple of seconds longer before sinking to my knees and declaring the *Thousand Shades of Green* challenge to be 'Mission accomplished.' Zachary whooped, Bramble looked at me quizzically and Christina just smiled, that broad, beautiful smile, before then saying 'Thank goodness for that.'

My sentiment exactly.

Acknowledgements

The list of folk who have offered advice, companionship and words of encouragement is not quite as long as the list of plants I managed to spot, but is lengthy nevertheless.

Firstly at Bloomsbury, I'd like to express my deepest appreciation to Jim Martin for commissioning such a crazy idea in the first place. My editor Jenny Campbell is so organised and has been fabulous at both keeping me on track throughout the whole process and offering words of encouragement at exactly the right time. Thanks also to both Elizabeth Peters for carrying out such a fine edit of the first draft and Jessica Gray and Sarah Head at Bloomsbury for ensuring the book reaches the widest possible readership. Thanks also to Kate Page for holding my hand during the recording of the audiobook – and making me laugh a lot.

The Somerset Rare Plants Group (SRPG) provided huge expertise and encouragement throughout the year and were fine company to boot. I must thank in particular Helena Crouch for her unwavering support and patience, despite a constant badgering from me with a thousand questions – her knowledge and dedication towards advancing the knowledge of Somerset's plants is second to none. Thanks also to Clive Lovatt who helped an enormous amount in the planning process, but sadly died before the book was published. I'm also indebted in particular to Cath Mowat, Liz McDonnell (who also sadly died), Ellen McDouall, Stephen Parker and Fred Rumsey – all SRPG stalwarts – for their immense knowledge and a shared passion for imparting it to those less knowledgeable while out in the field.

On early trips across to Bath, I enjoyed the company of my in-laws, Graham and Laura Holvey, while tracking down a fine array of woodland plants, and thanks also to my great naturalist friend Ed Drewitt for joining me on a grand day out to the Brecon Beacons. Hilary and Steve Pickersgill were most helpful during my time orchid-spotting in the Chilterns, and no one knows Gloucestershire's road verges better than Simon Harding. Across in Wiltshire, it was super to catch up with Ben Prater to talk about plants on the radio and thanks also to Sharon Pilkington for a tremendous crash course on aquatic plants in this much underrated county.

While down on the Lizard Peninsula, I was very grateful to Ian Bennallick and Gareth Jones for letting me gatecrash their clover party. I'm also deeply indebted to Simon Harrap, not only for guiding me around the botanical hotspots of the Brecks and North Norfolk, but also for letting me pick his fine naturalist's brain throughout the entirety of the year. During my numerous stays in Norfolk, I always have a welcome offer of accommodation at the home of Nigel Redman and Cheryle Sifontes, and it was particularly gratifying see Nigel botanising rather than birding – for once – during our time together!

I had the most enormous amount of fun plant-hunting with my old pal Tim Sykes down in the New Forest and would also like to extend my gratitude to Clive Chatters for pointing us in the right direction of the Forest's rarest plants. Up on Anglesey, Nigel Brown gave my mission the most enormous boost, and his wise council over the last 30 years continues to be a guiding light throughout my career. Thanks must also go to Caroline Brown for putting me up, and putting up with me during my numerous stays under the Brown roof over the years, and it was terrific to make the acquaintance of Wendy McCarthy, who knows the plants of the Great Orme like the back of her own hand. Tim Rich's kind permission to use an except of his and Wendy's paper on *Hieracium britannicoides* was gratefully received too. Also in west Wales, I delighted in the mad dash for Irish lady's-tresses with the TV Presenter Nigel Marven and his colleague Roger Harris, along with Nigel Redman again – who was also thrilled to tick off his last remaining British orchid.

A visit to Kent is essential for any plant-lister and I'm hugely indebted to Richard Moyse and Ben Sweeney for guiding me around Ranscombe Farm – surely one of the best plant reserves in England. While down in Kent, my old filming friend from BBC's *Nature's Calendar* and *The One Show*, Richard Taylor-Jones, also kindly took me in. While showing me his local patch, we also spent a wonderful day searching out rare broomrapes.

Visiting Teesdale was one of the highlights of my year and the trip was made immeasurably more rewarding thanks to the kind help and insider knowledge of Martin Furness of Natural England. North of the border, the Grant Arms Hotel in Grantown-on-Spey is my second home and thanks to all the staff for making me feel so welcome during my numerous stays over the course of the year. I consider Sue Williams and Simon Pawsey to be among my closest friends, and botanical expeditions to Golspie and Cairngorm, respectively were immeasurably enhanced by their first-rate company.

The expedition to Cùl Mòr with Barbara Jones was, for me, a career highlight. Barbara's expertise of upland ecology is immense and our joint delight at finding Norwegian mugwort while recording for BBC Radio 4's *Costing the Earth* is a moment I'll treasure. Thanks also to Barbara's husband, Nigel Jones, for being our point man while we were up a big hill in the middle of nowhere. Still up in Scotland, I'd like to thank Dave Jones for showing me his reserves and the Botanical Society of Britain and Ireland (BSBI) for organising two fabulous field trips to Glenshee and Ben Lawers on consecutive days. Here, Les Tucker, David Elston and Dan Watson took the assembled groups up hill and down dale, enabling me to see an astonishing 39 'lifers' during the course of one heady weekend. I also delighted in catching up with my new botanical buddy Mark Hows throughout the rest of the year, as we helped each other out with locations for a number of stellar rarities.

Starfruit was a plant that I was desperate to see from the outset and I'm thankful to Joanna Bromley for putting me in touch with the botanical guru that is Andy Byfield, who in turn was gracious enough to make it happen. A fleeting visit to Dorset to catch up with Miles King, also allowed us to not botanise on a nudist beach in Dorset and reminded me of some of the lasting friendships I've made during a career spent interviewing people for television.

The Wild Flower Society (WFS) run an array of marvellous field trips, thanks to Janet John and her team, and I'm indebted to Jackie Hardy, and John and Monika Walton for a fascinating day out in Worcestershire.

If you too fancy getting into wild plants, I highly recommend you join the BSBI, WFS and Plantlife charities. Their eagerness to help and support those keen to learn more about botany is nothing short of world class, and my modest membership fees have effectively been repaid many times over as a result of the help I've received.

I'm also extremely grateful to my mum, Renee and brothers Andy and Paul for always being so encouraging and supportive of my slightly precarious way of making a living. And finally, the biggest debt of gratitude goes to the Chew Stoke Botanical Group – aka my wife Christina, son Zachary and dog Bramble. Thank you, from the bottom of my heart, for indulging me yet again.

Gazetteer

Key: nr = near; FP = footpath; CP = Country Park; WT = Wildlife Trust; NT = National Trust; NNR = Nat Nature Reserve; LNR = Local Nature Reserve; SSSI = Site of Special Scientific Interest; WWT = Wild and Wetlands Trust; NR = Nature Reserve; N = North; S = South; E = East; W = West; NW = north- NE = north-east; SW = south-west; SE = south-east; WsM = Weston-super-Mare; H Wycombe = High Wyco Gloucs = Gloucestershire; Wilts = Wiltshire; CV = Chew Valley; GoS = Grantown-on-Spey; SWT = Somerset Wi Trust; Scottish WT = Scottish Wildlife Trust; BBOWT = Bucks, Berks and Oxon Wildlife Trust; NWWT = North V Wildlife Trust; GWT = Gloucestershire Wildlife Trust; CC = County Council; Bradford OA = Bradford-on-A Temp = temporary; UWE = University of the West of England; MOD = Ministry of Defence; NF = New F PL = Plantlife; RV = rendezvous.

No.	Date	Location	Plant species – common name	Plant species – scientific name
1	04-01-21	Chew Stoke to Chew Magna walk	Groundsel	*Senecio vulgaris*
2	04-01-21	Chew Stoke to Chew Magna walk	Lesser celandine	*Ranunculus ficaria*
3	04-01-21	Chew Stoke to Chew Magna walk	Dog's mercury	*Mercurialis perennis*
4	04-01-21	Chew Stoke to Chew Magna walk	Red dead-nettle	*Lamium purpureum*
5	04-01-21	Chew Stoke to Chew Magna walk	Common field speedwell	*Veronica persica*
6	04-01-21	Chew Stoke to Chew Magna walk	Scentless mayweed	*Tripleurospermum inodorum*
7	04-01-21	Chew Stoke to Chew Magna walk	Wild radish	*Raphanus raphanistru* ssp. *raphanistrum*
8	04-01-21	Chew Stoke to Chew Magna walk	Annual meadow grass	*Poa annua*
9	04-01-21	Chew Stoke to Chew Magna walk	Hogweed	*Heracleum sphondyliu*
10	04-01-21	Chew Stoke to Chew Magna walk	Petty spurge	*Euphorbia peplus*
11	04-01-21	Chew Stoke to Chew Magna walk	Common primrose	*Primula vulgaris*
12	04-01-21	Chew Stoke to Chew Magna walk	Wood avens	*Geum urbanum*
13	04-01-21	Chew Stoke to Chew Magna walk	Hairy bittercress	*Cardamine hirsuta*
14	04-01-21	Chew Stoke to Chew Magna walk	Ivy-leaved toadflax	*Cymbalaria muralis*
15	04-01-21	Chew Stoke to Chew Magna walk	Daisy	*Bellis perennis*
16	04-01-21	Chew Stoke to Chew Magna walk	Trailing bellflower	*Campanula poscharsk*
17	04-01-21	Chew Stoke to Chew Magna walk	Red valerian	*Centranthus ruber*
18	04-01-21	Chew Stoke to Chew Magna walk	Mexican fleabane	*Erigeron karvinskianu*
19	04-01-21	Chew Stoke to Chew Magna walk	Dandelion	*Taraxacum officinale* a
20	04-01-21	Chew Stoke to Chew Magna walk	Common ragwort	*Senecio jacobaea*
21	04-01-21	Chew Stoke to Chew Magna walk	Common evening primrose	*Oenothera biennis*
22	04-01-21	Chew Stoke to Chew Magna walk	Adria bellflower	*Campanula portenschlagiana*
23	04-01-21	Chew Stoke to Chew Magna walk	Canadian fleabane	*Erigeron canadensis*
24	04-01-21	Chew Stoke to Chew Magna walk	White dead-nettle	*Lamium album*
25	04-01-21	Chew Stoke to Chew Magna walk	Yarrow	*Achillea millefolium*
26	04-01-21	Chew Stoke to Chew Magna walk	Smooth hawksbeard	*Crepis capillaris*
27	04-01-21	Chew Stoke to Chew Magna walk	Smooth sow-thistle	*Sonchus oleraceus*
28	22-01-21	King's Castle Wood, nr Wells	Winter heliotrope	*Petasites pyrenaicus*
29	07-02-21	Weston Big Wood, nr Weston-in-Gordano	Small-leaved lime	*Tilia cordata*
30	07-02-21	Weston Big Wood, nr Weston-in-Gordano	Hart's tongue fern	*Asplenium scolopendri*
31	07-02-21	Weston Big Wood, nr Weston-in-Gordano	Holly	*Ilex aquifolium*
32	07-02-21	Weston Big Wood, nr Weston-in-Gordano	Lords and ladies	*Arum maculatum*
33	07-02-21	Weston Big Wood, nr Weston-in-Gordano	Elder	*Sambucus nigra*
34	07-02-21	Weston Big Wood, nr Weston-in-Gordano	Bramble	*Rubus fruticosis* agg.
35	07-02-21	Weston Big Wood, nr Weston-in-Gordano	Hawthorn	*Crataegus monogyna*
36	07-02-21	Weston Big Wood, nr Weston-in-Gordano	Blackthorn	*Prunus spinosa*
37	07-02-21	Weston Big Wood, nr Weston-in-Gordano	Spurge-laurel	*Daphne laureola*
38	07-02-21	Weston Big Wood, nr Weston-in-Gordano	Wood spurge	*Euphorbia amygdaloid*
39	07-02-21	Weston Big Wood, nr Weston-in-Gordano	Dogwood	*Cornus sanguinea*

. Date	Location	Plant species – common name	Plant species – scientific name
07-02-21	Weston Big Wood, nr Weston-in-Gordano	Hazel	*Corylus avellana*
07-02-21	Weston Big Wood, nr Weston-in-Gordano	Spindle	*Euonymus europeaus*
07-02-21	Weston Big Wood, nr Weston-in-Gordano	Wayfaring tree	*Viburnum lantana*
08-02-21	Road to yacht club, Chew Stoke	Snowdrop	*Galanthus nivalis*
11-02-21	Nr Salt & Malt, CV Lake	Shepherd's purse	*Capsella bursa-pastoris*
11-02-21	Nr Salt & Malt, CV Lake	Common reed	*Phragmites australis*
18-02-21	Ubley Warren Reserve, Somerset WT	Maidenhair spleenwort	*Asplenium trichomanes*
23-02-21	Herefordshire, various locations	Mistletoe	*Viscum album*
23-02-21	Stanner Rocks NNR, Powys	Broom	*Cystisus scoparius*
23-02-21	Stanner Rocks NNR, Powys	Navelwort	*Umbilicus rupestris*
23-02-21	Stanner Rocks NNR, Powys	Wood sage	*Teucrium scorodonia*
23-02-21	Stanner Rocks NNR, Powys	Wall speedwell	*Veronica arvensis*
23-02-21	Stanner Rocks NNR, Powys	Wall rue	*Asplenium ruta-muraria*
23-02-21	Stanner Rocks NNR, Powys	Black spleenwort	*Asplenium adiantum-nigrum*
23-02-21	Stanner Rocks NNR, Powys	Early star-of-Bethlehem	*Gagea bohemica*
26-02-21	Ubley Warren Reserve, Somerset WT	Soft shield-fern	*Polystichum setiferum*
01-03-21	Highgate Wood, Muswell Hill	Wild service tree	*Sorbus torminalis*
01-03-21	Highgate Wood, Muswell Hill	Hornbeam	*Carpinus betulinus*
14-03-21	Breach Hill to Gravel Hill, Chew Stoke	Early dog violet	*Viola reichenbachiana*
14-03-21	Breach Hill to Gravel Hill, Chew Stoke	Barren strawberry	*Potentilla sterilis*
14-03-21	Breach Hill to Gravel Hill, Chew Stoke	Sweet violet	*Viola odorata*
14-03-21	Breach Hill to Gravel Hill, Chew Stoke	Hairy violet	*Viola hirta*
14-03-21	School Lane, Chew Stoke	Rustyback fern	*Asplenium ceterach*
17-03-21	Create Centre, Smeaton Road, Bristol	Alexanders	*Smyrnium olusatrum*
17-03-21	Create Centre, Smeaton Road, Bristol	Three-cornered garlic	*Allium triquetum*
17-03-21	Cumberland Road, Bristol	Common whitlowgrass	*Erophila verna*
17-03-21	Cumberland Road, Bristol	Rue-leaved saxifrage	*Saxifraga tridactylites*
17-03-21	Cumberland Road, Bristol	Thale cress	*Arabidopsis thaliana*
17-03-21	Avon Street, Temple Quay, Bristol	Coltsfoot	*Tussilago farfara*
20-03-21	Wallycourt Road, Chew Stoke	Ground Ivy	*Glechoma hederacea*
20-03-21	The Community Farm, Chew Stoke	Cowslip	*Primula veris*
21-03-21	Freshford, nr Bath	Wood anemone	*Anemone nemorosa*
21-03-21	Freshford, nr Bath	Wild daffodil	*Narcissus pseudonarcissus*
21-03-21	Freshford, nr Bath	Toothwort	*Lathraea squamaria*
26-03-21	Path to Fisherman's car park, Chew Stoke	Russian comfrey	*Symphytum x uplandicum*
28-03-21	Ubley Warren Reserve, Somerset WT	Hutchinsia	*Hornungia petraea*
28-03-21	Combe Lane, N of Charlton Adam	Perfoliate (Cotswold) pennycress	*Thlaspi perfoliatum*
29-03-21	Michael Wood Services slip-road, M5 N	Danish scurvy-grass	*Cochlearia danica*
31-03-21	Rectory Fields, Chew Stoke	Slender speedwell	*Veronica filiformis*
02-04-21	Freshford, nr Bath	Shining cranesbill	*Geranium lucidum*
02-04-21	Freshford, nr Bath	Cuckooflower	*Cardamine pratensis*
02-04-21	Freshford, nr Bath	Common dog violet	*Viola riviniana*
02-04-21	Freshford Village, nr Bath	Yellow corydalis	*Pseudofumaria lutea*
02-04-21	Freshford Village, nr Bath	Pellitory-of-the-wall	*Parietaria judaica*
02-04-21	Freshford Village, nr Bath	Herb robert	*Geranium robertianum*
02-04-21	Freshford, nr Bath	Opposite-leaved golden saxifrage	*Chrysosplenum oppositifolium*
04-04-21	East Harptree Woods, nr East Harptree	Wood sorrel	*Oxalis acetosella*
04-04-21	East Harptree Woods, nr East Harptree	Wild raspberry	*Rubus idaeus*
04-04-21	Lily Combe, W of Chewton Mendip	Green hellebore	*Helleborus viridus*
05-04-21	Disused quarries, Stoke St Michael	Field forget-me-not	*Myosotis arvensis*
05-04-21	Brownfield site on A36 nr Costa, Bath	Oxford ragwort	*Senecio squalidus*
05-04-21	Brownfield site on A36 nr Costa, Bath	Butterfly bush	*Buddleja davidii*
05-04-21	St John's Road, nr Victoria Park, Bath	Keeled-fruited cornsalad	*Valerianella carinata*
05-04-21	Upper Bristol Road, nr Victoria Park, Bath	Sticky mouse-ear	*Cerastium glomeratum*

No.	Date	Location	Plant species - common name	Plant species - scientific name
94	06-04-21	Bithams Woods, nr Chew Magna	Greater stitchwort	*Stellaria holostea*
95	06-04-21	Bithams Woods, nr Chew Magna	Moschatel	*Adoxa moschatellina*
96	07-04-21	Dawlish Warren NNR, nr visitor centre	Slender parsleypiert	*Aphanes australis*
97	07-04-21	Dawlish Warren NNR, nr visitor centre	Early forget-me-not	*Myosotis ramosissima*
98	07-04-21	Dawlish Warren NNR, nr visitor centre	Shepherd's cress	*Teesdalia nudicaulis*
99	07-04-21	Dawlish Warren NNR, nr visitor centre	Sand crocus	*Romulea columnae*
100	07-04-21	Exmouth Esplanade, Exmouth	Common ramping fumitory	*Fumaria muralis*
101	11-04-21	Black Rock NR, nr Cheddar	Lesser trefoil	*Trifolium dubium*
102	11-04-21	Black Rock NR, nr Cheddar	Common storksbill	*Erodium cicutarium*
103	11-04-21	Black Rock NR, nr Cheddar	Spring cinquefoil	*Potentilla verna*
104	11-04-21	Velvet Bottom, Somerset WT	Dwarf mouse-ear	*Cerastium pumilum*
105	11-04-21	Velvet Bottom, Somerset WT	Water horsetail	*Equisetum fluviatile*
106	11-04-21	Velvet Bottom, Somerset WT	Marsh marigold	*Caltha palustris*
107	11-04-21	High Street, Cheddar	Green alkanet	*Pentaglottis sempervi*
108	11-04-21	High Street, Cheddar	Wild strawberry	*Fragaria vescans*
109	13-04-21	Goblin Combe NR, Bristol	Ribwort plantain	*Plantago lanceolata*
110	14-04-21	Behind Bristol Water, nr CV Lake	Wych elm	*Ulmus glabra*
111	14-04-21	Wally Lane, nr CV Lake	Spring sedge	*Carex caryophyllea*
112	14-04-21	Wally Lane, nr CV Lake	Meadow foxtail	*Alopecuris pratensis*
113	14-04-21	Portway, nr Clifton Suspension Bridge	Wallflower	*Erysimum cheiri*
114	14-04-21	N of Observatory, Avon Gorge	Creeping cinquefoil	*Potentilla reptans*
115	14-04-21	N of Observatory, Avon Gorge	Bristol rock-cress	*Arabis scabra*
116	14-04-21	Gully, by Circular road, Clifton Downs	Salad burnet	*Poterium sanguisorba*
117	14-04-21	Gully, by Circular road, Clifton Downs	Common rockrose	*Helianthemum nummularium*
118	15-04-21	Gully, by Circular road, Clifton Downs	Honewort	*Trinia glauca*
119	15-04-21	Stow Road, 3 miles, N of Cirencester	Pasqueflower	*Pulsatilla vulgaris*
120	15-04-21	North Meadow NNR, N of Cricklade	Snake's-head fritillary	*Fritillaria meleagris*
121	15-04-21	Roadside by North Meadow NNR	Butterbur	*Petasites hybridus*
122	15-04-21	Roadside by North Meadow NNR	Garlic mustard	*Alliaria petiolata*
123	16-04-21	S end of Kennet and Avon Canal, Bath	Wintercress	*Barbarea vulgaris*
124	16-04-21	S end of Kennet and Avon Canal, Bath	Wavy bittercress	*Cardamine flexuosa*
125	20-04-21	Randolph Ave, Yate, Bristol	Thyme-leaved speedwell	*Veronica serpyllifolia*
126	20-04-21	Hursley Lane, off A37, S of Whitchurch	Honesty	*Lunaria annua*
127	23-04-21	Breach Hill Lane, Chew Stoke	Cow parsley	*Anthriscus sylvestris*
128	23-04-21	Breach Hill Lane, Chew Stoke	Ramsons	*Allium ursinum*
129	23-04-21	Breach Hill Lane, Chew Stoke	Bush vetch	*Vicia sepium*
130	23-04-21	Breach Hill Lane, Chew Stoke	Bluebell	*Hyacinthoides non-sc*
131	23-04-21	Kingshill Lane, Chew Stoke	Yellow archangel	*Lamiastrum galeobdo*
132	23-04-21	Kingshill Lane, Chew Stoke	Bulbous buttercup	*Ranunculus bulbosus*
133	23-04-21	Kingshill Lane, Chew Stoke	Germander speedwell	*Veronica chamaedrys*
134	23-04-21	B3114, W of CV Lake	False oxlip	*Primula x polyantha*
135	23-04-21	B3114, W of CV Lake	Bugle	*Ajuga reptans*
136	23-04-21	Bushythorn Road, Chew Stoke	Charlock	*Sinapsis arvensis*
137	24-04-21	Craig Cerrig Gleisiad NNR, Brecon Beacons	Purple saxifrage	*Saxifraga oppositifolia*
138	24-04-21	Dundry Hill, S of Bristol	Red campion	*Silene dioica*
139	26-04-21	Rectory Fields, Chew Stoke	Creeping buttercup	*Ranunculus repens*
140	28-04-21	Fields E of Chota Castle, Chew Magna	Common sorrel	*Rumex acetosa*
141	28-04-21	Fields E of Chota Castle, Chew Magna	Meadow buttercup	*Ranunculus acris*
142	28-04-21	Fields E of Chota Castle, Chew Magna	Red clover	*Trifolium pratense*
143	29-04-21	Greenditch Lane, N of Chilcompton	Wall whitlowgrass	*Draba muralis*
144	29-04-21	Velvet Bottom, Somerset WT	Alpine pennycress	*Noccaea caerulescens*
145	29-04-21	Parsonage Lane, Cheddar Woods, nr Axbridge	Wood melic	*Melica uniflora*
146	29-04-21	Parsonage Lane, Cheddar Woods, nr Axbridge	Goldilocks buttercup	*Ranunculus auricomu*
147	29-04-21	Grassland NW corner, Cheddar Woods	Early-purple orchid	*Orchis mascula*

o.	Date	Location	Plant species - common name	Plant species - scientific name
8	29-04-21	Grassland NW corner, Cheddar Woods	Mouse-ear hawkweed	*Pilosella officinarum*
9	29-04-21	Grassland NW corner, Cheddar Woods	Parsley-piert	*Aphanes arvensis*
0	29-04-21	E FP, Cheddar Woods	Purple gromwell	*Aegonychon purpureocaeruleum*
1	29-04-21	Nr A371, Cheddar Woods	Butcher's broom	*Ruscus aculeatus*
2	29-04-21	Nr A371, Cheddar Woods	Cock's-foot grass	*Dactylus glomerata*
3	29-04-21	Path to Sand Point NT, N of WsM	Pink sorrel	*Oxalis articulata*
4	29-04-21	Path to Sand Point NT, N of WsM	Dove's foot cranesbill	*Geranium molle*
5	29-04-21	Top of Sand Point NT, N of WsM	Bird's-foot trefoil	*Lotus corniculatus*
6	29-04-21	W side of Sand Point NT	Common vetch	*Vicia sativa*
7	29-04-21	W side of Sand Point NT	Sea campion	*Silene uniflora*
8	29-04-21	Top of Sand Point NT	Suffocated clover	*Trifolium suffocatum*
9	29-04-21	Top of stairs, Sand Point NT	Wild clary	*Salvia verbenaca*
0	29-04-21	Top of stairs, Sand Point NT	Field madder	*Sherardia arvensis*
1	29-04-21	Beach Road, Sand Point NT	Greater celandine	*Chelidonium majus*
2	30-04-21	Sandy Lane, Stanton Drew	Barren brome	*Anisantha sterilis*
3	01-05-21	Walk to Chew Magna	Cut-leaved cranesbill	*Geranium dissectum*
4	02-05-21	Bay Road, Ladye Bay, Clevedon	White bryony	*Bryonia dioica*
5	02-05-21	Steps down to beach, Ladye Bay, Clevedon	Brookweed	*Samolus valerandi*
6	02-05-21	Path above the beach, Ladye Bay	Wild madder	*Rubia peregrina*
7	02-05-21	Path above the beach, Ladye Bay	Three-veined sandwort	*Moehringia trinerva*
8	02-05-21	Path above the beach, Ladye Bay	Rough sow-thistle	*Sonchus asper*
9	02-05-21	Path above the beach, Ladye Bay	Cleavers	*Galium aparine*
0	02-05-21	Path above the beach, Ladye Bay	Broad buckler fern	*Dryopteris dilitata*
1	02-05-21	Nr sea, Ladye Bay, Clevedon	Sea spleenwort	*Asplenium marinum*
2	02-05-21	On rock platform, nr sea, Clevedon	Buckshorn plantain	*Plantago coronopus*
3	02-05-21	On top of a rock, by sea, Clevedon	Rock samphire	*Crithmum maritimum*
4	02-05-21	On top of rock platform, by sea	Sea pearlwort	*Sagina maritima*
5	02-05-21	On top of rock, vegetated zone	Distant sedge	*Carex distans*
6	02-05-21	Short grassland, by sea, Clevedon	Glaucous sedge	*Carex flacca*
7	02-05-21	Along FP away from sea, Clevedon	Common chickweed	*Stellaria media*
8	02-05-21	Clevedon Golf Course, Clevedon	Thick-leaved stonecrop	*Sedum dasyphyllum*
9	02-05-21	Clevedon Golf Course, Clevedon	Swine-cress	*Lepidium coronopus*
0	02-05-21	Clevedon Golf Course, Clevedon	Annual mercury	*Mercurialis annua*
1	02-05-21	Clevedon Golf Course, Clevedon	Black mustard	*Brassica nigra*
2	02-05-21	Castle Road, Clevedon	Thyme-leaved sandwort	*Arenaria serpyllifolia*
3	02-05-21	Castle Road, Clevedon	Soft brome	*Bromus hordeaceous*
4	02-05-21	Island in road, Castle Road, Clevedon	Procumbent pearlwort	*Sagina procumbens*
5	02-05-21	Jn of Castle Road and Bay Road, Clevedon	California poppy	*Eschsolzia californica*
6	03-05-21	S end of Sand Bay Beach, WsM	Sea spurge	*Euphorbia paralias*
7	03-05-21	S end of Sand Bay Beach, WsM	Sand sedge	*Carex arenaria*
8	03-05-21	S/middle of Sand Bay Beach, WsM	Sea sandwort	*Honckenya peploides*
9	03-05-21	S/middle of Sand Bay Beach, WsM	Beaked hawksbeard	*Crepis vesicaria*
0	03-05-21	S/middle of Sand Bay Beach, WsM	Sea beet	*Beta vulgaris* ssp. *maritima*
1	03-05-21	On rocks, S of beach, WsM	Common scurvy-grass	*Cochlearia officinalis*
2	03-05-21	Between road and dunes, Sand Bay, WsM	Hairy tare	*Ervilia hirsuta*
3	03-05-21	On rocks, S of beach, WsM	Sheep's fescue	*Festuca ovina*
4	05-05-21	Freda's Grave, Cannock Chase CP, Staffs	Heather	*Calluna vulgaris*
5	05-05-21	Freda's Grave, Cannock Chase CP, Staffs	Cowberry	*Vaccinium vitis-idaea*
6	05-05-21	Freda's Grave, Cannock Chase CP, Staffs	Bilberry	*Vaccinium myrtillus*
7	06-05-21	Hollow Marsh, Somerset WT reserve	Solomon's seal	*Polygonatum multiflorum*
8	06-05-21	Hollow Marsh, Somerset WT reserve	Sweet woodruff	*Galium odoratum*
9	06-05-21	Hollow Marsh, Somerset WT reserve	Bitter-vetch	*Lathyrus linifolius*
0	06-05-21	Hollow Marsh, Somerset WT reserve	Wood sedge	*Carex sylvatica*
1	06-05-21	On roadside by salt storage dome, A37	Hoary cress	*Lepidium draba*
2	08-05-21	Nunnery Copse Fields, CV Lake	Green-winged orchid	*Anacamptis morio*

No.	Date	Location	Plant species - common name	Plant species - scientific name
203	08-05-21	Nunnery Copse Fields, CV Lake	Silverweed	*Potentilla anserina*
204	08-05-21	Nunnery Copse Fields, CV Lake	Upright brome	*Bromopsis erecta*
205	08-05-21	Nunnery Copse Fields, CV Lake	Black medick	*Medicago lupulina*
206	09-05-21	Top of point, Sand Point NT	Spotted medick	*Medicago arabica*
207	09-05-21	Top of point, Sand Point NT	Changing forget-me-not	*Myosotis discolour*
208	09-05-21	Top of point, Sand Point NT	Lesser chickweed	*Stellaria pallida*
209	09-05-21	Grassland behind dunes Point, Sand Point NT	Red valerian	*Centranthus ruber*
210	09-05-21	Grassland behind dunes Point, Sand Point NT	Drooping star-of-Bethlehem	*Ornithogalum nutans*
211	09-05-21	Grassland behind dunes Point, Sand Point NT	Sweet Alison	*Lobularia maritima*
212	10-05-21	Wavering Down, Crook Peak, Mendips	Fairy flax	*Linum catharticum*
213	11-05-21	Nr Quantocks Hide, WWT Steart Marshes	Ragged robin	*Lychnis flos-cuculi*
214	11-05-21	Car park, WWT Steart Marshes	Long-stalked cranesbill	*Geranium columbinum*
215	11-05-21	Saltmarsh, River Parrett Trail, WWT Steart	Marsh seagrass	*Triglochin maritima*
216	11-05-21	Saltmarsh, River Parrett Trail, WWT Steart	Sea purslane	*Atriplex portulacoides*
217	11-05-21	Saltmarsh, River Parrett Trail, WWT Steart	English scurvy-grass	*Cochlearia anglica*
218	11-05-21	Jack's drove, Tealham Moor, Somerset Levels	Celery-leaved buttercup	*Ranunculus scleratus*
219	15-05-21	Hay Wood, S of Hutton	Wood speedwell	*Veronica montana*
220	15-05-21	Hay Wood, S of Hutton	Caucasian saxifrage	*Saxifraga cymbalaria*
221	17-05-21	Minor road NE of Hartslock BBOWT NR	Hedgerow cranesbill	*Geranium pyrenaicum*
222	17-05-21	W of Hartslock NR	Common milkwort	*Polygala vulgaris*
223	17-05-21	On slope up to main orchid site, Hartslock NR	Common twayblade	*Neottia ovata*
224	17-05-21	By top of site, Hartslock NR	Monkey x lady orchid	*Orchis angusticrurus*
225	17-05-21	Further down slope, Hartslock NR	Monkey orchid	*Orchis simia*
226	17-05-21	Rapeseed field, minor road NE of Hartslock	Field pansy	*Viola arvensis*
227	17-05-21	Rapeseed field, minor road NE of Hartslock	Hedge mustard	*Sisymbrium officinale*
228	17-05-21	Under beech trees, Warburg NR BBOWT	Sanicle	*Sanicula europaea*
229	17-05-21	Along bridleway, Warburg NR BBOWT	Southern wood-rush	*Luzula forsteri*
230	17-05-21	By River Loddon, Dinton Pastures CP	Loddon lily	*Leucojum aestivum*
231	18-05-21	E end of meadow, Homefield Wood NR	Military orchid	*Orchis militaris*
232	18-05-21	E end of meadow, Homefield Wood NR	Fly orchid	*Ophrys insectifera*
233	18-05-21	W of Meadow, under beech, Homefield NR	Bird's-nest orchid	*Neottia nidus-avis*
234	18-05-21	Gomm's Wood, off Cock Lane, H Wycombe	Crosswort	*Cruciata laevipes*
235	18-05-21	Gomm's Wood, off Cock Lane, H Wycombe	Wood millet	*Millium effusum*
236	18-05-21	Gomms and Bubbles Wood, nr Hughendon	Coralroot	*Cardamine bulbifera*
237	19-05-21	RSPB Ham Wall, Somerset Levels	Yellow flag	*Iris pseudacorus*
238	20-05-21	Sweet Track, Shapwick NNR, Levels	Royal fern	*Osmunda regalis*
239	21-05-21	Car park, Westhay Moor NR, Somerset WT	Lesser swine-cress	*Lepidium didymum*
240	22-05-21	Car park, Westhay Moor NR, Somerset WT	Perennial ryegrass	*Lolium perenne*
241	23-05-21	Raised mire, Westhay Moor NR, Somerset WT	Sheep's sorrel	*Rumes acetosella*
242	24-05-21	Raised mire, Westhay Moor NR, Somerset WT	Common cotton-sedge	*Eriophorum angustifoli*
243	25-05-21	Raised mire, Westhay Moor NR, Somerset WT	Round-leaved sundew	*Drosera rotundifolia*
244	26-05-21	Chew River, S of Chew Magna	Meadow saxifrage	*Saxifraga granulata*
245	27-05-21	Beech plantation, Badminton Estate, A46	White helleborine	*Cephalanthera damasonium*
246	28-05-21	Temp Pond, Inglestone Common, Gloucs	Water mint	*Mentha aquatica*
247	29-05-21	Temp Pond, Inglestone Common, Gloucs	Lesser spearwort	*Ranunculus flammula*
248	30-05-21	Temp Pond, Inglestone Common, Gloucs	Thread-leaved water-crowfoot	*Ranunculus tricophyllu*
249	31-05-21	Temp Pond, Inglestone Common, Gloucs	Brooklime	*Veronica beccabunga*
250	01-06-21	Roadside verge, Frogland Cross, Gloucs	Bithynian vetch	*Vicia bithynica*
251	02-06-21	Roadside verge, Frogland Cross, Gloucs	Oxeye daisy	*Leucanthemum vulgare*
252	03-06-21	Tockington Hill, nr Olveston, Gloucs	Sicillian honey garlic	*Nectaroscordum siculum*
253	28-05-21	Leigh Woods, N Somerset	Yellow pimpernel	*Lysimachia nemorum*
254	28-05-21	Leigh Woods, Stoke Camp, N Somerset	Tutsan	*Hypericum androsaemu*
255	28-05-21	Leigh Woods, N Somerset	Columbine	*Aquilegia vulgaris*

No.	Date	Location	Plant species - common name	Plant species - scientific name
56	28-05-21	Leigh Woods, Stoke Camp, N Somerset	Rough hawksbeard	*Crepis biennis*
57	28-05-21	Leigh Woods, SW of Stoke Camp	Common spotted orchid	*Dactylorhiza fuchsii*
58	28-05-21	Leigh Woods, SW of Stoke Camp	Smooth meadowgrass	*Poa pratensis*
59	28-05-21	Leigh Woods, SW of Stoke Camp	Pignut	*Conopodium majus*
60	28-05-21	Leigh Woods, SW of Stoke Camp	English whitebeam	*Sorbus anglica*
61	28-05-21	Leigh Woods, SW of Stoke Camp	Bristol whitebeam	*Sorbus bristoliensis*
62	28-05-21	Leigh Woods, SW of Stoke Camp	Wilmott's whitebeam	*Sorbus wilmottiana*
63	30-05-21	Long Wood, Somerset WT	Herb-paris	*Paris quadrifolia*
64	30-05-21	Westbury Road, Westbury-on-Trym	Nipplewort	*Lapsana communis*
65	31-05-21	W of Red River, Marazion Beach	Seaside daisy	*Erigeron glaucus*
66	31-05-21	W of Red River, Marazion Beach	Sea holly	*Eryngium maritimum*
67	31-05-21	In dunes, Marazion Beach, Cornwall	Kidney vetch	*Anthyllis vulneraria*
68	31-05-21	In dunes, Marazion Beach, Cornwall	Japanese rose	*Rosa rugosa*
69	31-05-21	In dunes, Marazion Beach, Cornwall	Sea daffodil	*Pancratium maritimum*
70	01-06-21	In dunes, Marazion Beach, Cornwall	Sea bindweed	*Calystegia soldanella*
71	01-06-21	N of Little Trevothan, Lizard Peninsula	Eastern gladiolus	*Gladiolus communis* ssp. *byzantinus*
72	01-06-21	Dog-walking area, Little Trevothan, Lizard	Hemlock water dropwort	*Oenanthe crocata*
73	01-06-21	Dog-walking area, Little Trevothan, Lizard	Foxglove	*Digitalis purpurea*
74	01-06-21	Dog-walking area, Little Trevothan, Lizard	Yorkshire fog	*Holcus lanatus*
75	01-06-21	Dog-walking area, Little Trevothan, Lizard	Water figwort	*Scrophularia auriculata*
76	01-06-21	NT car park, Kynance Cove, Lizard	Thrift	*Armeria maritima*
77	01-06-21	Coastal walk to Kynance Cove, Lizard	Bloody cranesbill	*Geranium sanguineum*
78	01-06-21	Coastal walk to Kynance Cove, Lizard	Wild carrot	*Daucus carota*
79	01-06-21	Steep slope N of Kynance Cove, Lizard	Spring sandwort	*Minuartia verna*
80	01-06-21	Steep slope N of Kynance Cove, Lizard	Sheepsbit	*Jasione montana*
81	01-06-21	Steep slope N of Kynance Cove, Lizard	Catsear	*Hypochaeris radicata*
82	01-06-21	Steep slope N of Kynance Cove, Lizard	Sea mayweed	*Tripleurospermum maritime*
83	01-06-21	Steep slope N of Kynance Cove, Lizard	Fringed rupturewort	*Herniaria ciliolata*
84	01-06-21	Steep slope N of Kynance Cove, Lizard	Heath milkwort	*Polygala serpyllifolia*
85	01-06-21	Steep slope N of Kynance Cove, Lizard	Hairy greenweed	*Genista pilosa*
86	01-06-21	By steps nr Kynance Cove Cafe, Lizard	Tree mallow	*Malva arborea*
87	01-06-21	Steep slope N of Kynance Cove, Lizard	Spring squill	*Scilla verna*
88	01-06-21	Steep slope N of Kynance Cove, Lizard	Petty whin	*Genista anglica*
89	01-06-21	Steep slope N of Kynance Cove, Lizard	Wild thyme	*Thymus praecox*
90	01-06-21	Steep slope N of Kynance Cove, Lizard	Spotted catsear	*Hypochaeris maculata*
91	01-06-21	Steep slope N of Kynance Cove, Lizard	Dropwort	*Filipendula vulgaris*
92	01-06-21	Steep slope N of Kynance Cove, Lizard	Lousewort	*Pedicularis sylvatica*
93	01-06-21	Coastal path, N of Kynance Cove, Lizard	Black bog-rush	*Schoenus nigricans*
94	01-06-21	Coastal path, N of Kynance Cove, Lizard	Heath-spotted orchid	*Dactylorhiza maculata*
95	01-06-21	Coastal path, N of Kynance Cove, Lizard	Eyebright	*Euphrasia* spp.
96	01-06-21	Coastal path, N of Kynance Cove, Lizard	White stonecrop	*Sedum album*
97	01-06-21	Zigzag path up to NT car park, Lizard	Honeysuckle	*Lonicera periclymenum*
98	02-06-21	Wall to Gunwalloe Cove Beach, Lizard	Rock sea-spurrey	*Spergularia rupicola*
99	02-06-21	By church wall, Gunwalloe Cove Beach	Sea rocket	*Cakile maritima*
00	02-06-21	Road out of Gunwalloe Cove, Lizard	Babington's leek	*Allium ampeloprasum* var. *babingtonii*
01	02-06-21	Roadside, just W of Coverack, Lizard	Rhododendron	*Rhododendron ponticum*
02	02-06-21	Roadside verge in Coverack, Lizard	Scarlet pimpernel	*Anagallis arvensis*
03	02-06-21	Dog-walking area, Little Trevothan, Lizard	Toad rush	*Juncus bufonius* agg.
04	02-06-21	Croft Pascoe Pool, The Lizard NNR	Common spike-rush	*Eleocharis lacustris*
05	03-06-21	Path to Caerthillian Cove, Lizard	Wall barley	*Hordeum murinum*
06	03-06-21	Path to Caerthillian Cove, Lizard	Common poppy	*Papaver rhoeas*
07	03-06-21	Path to Caerthillian Cove, Lizard	Greater quaking grass	*Briza maxima*
08	03-06-21	S of path, Caerthillian Cove, Lizard	Upright clover	*Trifolium strictum*
09	03-06-21	S of path, Caerthillian Cove, Lizard	Rough clover	*Trifolium scabram*

No.	Date	Location	Plant species - common name	Plant species - scientific name
310	03-06-21	S of path, Caerthillian Cove, Lizard	Burrowing clover	*Trifolium subterraneum*
311	03-06-21	S of path, Caerthillian Cove, Lizard	Long-headed clover	*Trifolium incarnatum*
312	03-06-21	Lizard Point, Lizard Peninsula	Hottentot-fig	*Carpobrotus edulis*
313	05-06-21	Lizard Point, Lizard Peninsula	Purple dewplant	*Disphyma crassifolium*
314	03-06-21	By harbour, Mullion Cove, Lizard	Lesser sea-spurrey	*Spergularia marina*
315	03-06-21	By harbour, Mullion Cove, Lizard	Sea fern grass	*Catapodium marine*
316	04-06-21	S of path, Caerthillian Cove, Lizard	Western clover	*Trifolium occidentale*
317	04-06-21	S of path, Caerthillian Cove, Lizard	Knotted clover	*Trifolium striatum*
318	04-06-21	On path from Lizard to Caerthillian, Lizard	White clover	*Trifolium repens*
319	04-06-21	Croft Pasco Pool, The Lizard NNR	Bog pondweed	*Potamogeton polygonifo*
320	04-06-21	Croft Pasco Pool, The Lizard NNR	Soft rush	*Juncus effusus*
321	04-06-21	Along road in Coverack, Lizard	Pineappleweed	*Matricaria discoidea*
322	05-06-21	By road, FP and bridleway, Traboe, Lizard	Flea sedge	*Carex pulicaria*
323	05-06-21	By road, FP and bridleway, Traboe, Lizard	Heath dog violet	*Viola canina*
324	05-06-21	By road, FP and bridleway, Traboe, Lizard	Pale dog violet	*Viola lactea*
325	05-06-21	By road, FP and bridleway, Traboe, Lizard	Cornish heath	*Erica vagans*
326	05-06-21	By road, FP and bridleway, Traboe, Lizard	Common yellow sedge	*Carex demissa*
327	05-06-21	By road, FP and bridleway, Traboe, Lizard	Chives	*Allium schoenoprasum*
328	05-06-21	By road, FP and bridleway, Traboe, Lizard	Sea plantain	*Plantago maritima*
329	05-06-21	By road, FP and bridleway, Traboe, Lizard	Black sedge	*Carex nigra*
330	05-06-21	Nr Heliport, A30, nr Penzance	Common mallow	*Malva sylvestris*
331	05-06-21	Along stream, Nanquidno Valley, nr St Just	Giant rhubarb	*Gunnera tinctoria*
332	05-06-21	SW Coast Path, Nanquidno Valley	Bell heather	*Erica cinerea*
333	05-06-21	SW Coast Path, Nanquidno Valley	Fine-leaved sheep's fescue	*Festuca filiformis*
334	05-06-21	FP to Boscregan, N of Nanquidno	Spear thistle	*Cirsium vulgare*
335	05-06-21	FP to Boscregan, N of Nanquidno	Dodder	*Cuscuta epythymum*
336	05-06-21	3 fields at top of hill, Boscregan Farm NT	Purple viper's bugloss	*Echium plantagineum*
337	05-06-21	By stile entrance, Boscregan Farm NT	Slender trefoil	*Trifolium macranthum*
338	05-06-21	SW Coast Path, Nanquidno Valley	Spear-leaved orache	*Atriplex prostrata*
339	07-06-21	Cranwich Heath, Breckland	White campion	*Silene latifolia*
340	07-06-21	Cranwich Heath, Breckland	Hairy rock-cress	*Arabis hirsuta*
341	07-06-21	Cranwich Heath, Breckland	Smooth rupturewort	*Herniaria glabra*
342	07-06-21	Cranwich Heath, Breckland	Fine-leaved sandwort	*Minuartia hybrida*
343	07-06-21	Cranwich Heath, Breckland	Hop trefoil	*Trifolium campestre*
344	07-06-21	Cranwich Heath, Breckland	Purple milk-vetch	*Astragalus danicus*
345	07-06-21	Cranwich Heath, Breckland	Purple-stem cat's-tail	*Phleum phleoides*
346	07-06-21	Cranwich Heath, Breckland	Crested hair-grass	*Koeleria macrantha*
347	07-06-21	Cranwich Heath, Breckland	Bur medick	*Medicago minima*
348	07-06-21	Cranwich Heath, Breckland	Field mouse-ear	*Cerastium arvense*
349	07-06-21	Cranwich Heath, Breckland	Oregon grape	*Mahonia aquifolium*
350	07-06-21	Santon Downham, Breckland	Bird's-foot	*Ornithopus perpusillus*
351	07-06-21	Santon Downham, Breckland	Heath bedstraw	*Galium saxatile*
352	07-06-21	Santon Downham, Breckland	Heath groundsel	*Senecio sylvaticus*
353	07-06-21	Santon Downham, Breckland	Smooth catsear	*Hypochaeris glabra*
354	07-06-21	Santon Downham, Breckland	Annual knawel	*Scleranthus annuus*
355	07-06-21	Santon Downham, Breckland	Small cudweed	*Filago minima*
356	07-06-21	Santon Downham, Breckland	Hoary cinquefoil	*Potentilla argentea*
357	07-06-21	Santon Downham, Breckland	Clustered clover	*Trifolium glomeratum*
358	07-06-21	Santon Downham, Breckland	Corn spurrey	*Spergula arvensis*
359	07-06-21	Santon Downham, Breckland	Early hair-grass	*Aira praecox*
360	07-06-21	Santon Downham, Breckland	Wavy hair-grass	*Deschampsa flexuosa*
361	07-06-21	Santon Warren, Breckland	Perennial knawel	*Scleranthus perennis*
362	07-06-21	Santon Warren, Breckland	Tower mustard	*Arabis glabra*
363	07-06-21	Santon Warren, Breckland	Wild mignonette	*Reseda lutea*
364	07-06-21	Santon Warren, Breckland	Biting stonecrop	*Sedum acre*

No.	Date	Location	Plant species - common name	Plant species - scientific name
65	07-06-21	Santon Warren, Breckland	Mossy stonecrop	*Crassula tillaea*
66	07-06-21	By rail crossing, NE of Santon Downham	Smith's pepperwort	*Lepidium heterophyllum*
67	07-06-21	Brandon NR, Industrial Estate, Brandon	Field wormwood	*Artemisia campestris*
68	07-06-21	Brandon NR, Industrial Estate, Brandon	Hairy sedge	*Carex hirta*
69	07-06-21	Brandon NR, Industrial Estate, Brandon	Sickle Medick	*Medicago sativa* ssp. *falcata*
70	07-06-21	Cranwich Camp, Breckland	Caper spurge	*Euphorbia lathyris*
71	07-06-21	Cranwich Camp, Breckland	Spanish catchfly	*Silene otites*
72	08-06-21	Santon Warren, Breckland	Hare's-foot clover	*Trifolium arvense*
73	09-06-21	A149 between Salthouse and Kelling, Norfolk	Bugloss	*Anchusa arvensis*
74	09-06-21	A149 between Salthouse and Kelling, Norfolk	White campion	*Silene latifolia*
75	09-06-21	A149 between Salthouse and Kelling, Norfolk	Bladder campion	*Silene vulgaris*
76	09-06-21	Little Eye, Beach Road, Salthouse, Norfolk	Tree lupin	*Lupinus arboreus*
77	09-06-21	Shingle ridge at Salthouse, Norfolk	Curled dock	*Rumex crispus*
78	09-06-21	Shingle ridge at Salthouse, Norfolk	Yellow-horned poppy	*Glaucium flavum*
79	09-06-21	Car-park ditch, Salthouse, Norfolk	Sea club-rush	*Bolboschoenus maritimus*
80	09-06-21	Beeston Bump, nr Sheringham	Greater periwinkle	*Vinca major*
81	09-06-21	Beeston Bump, nr Sheringham	White lupin	*Lupinus albus*
82	09-06-21	Beeston Bump, nr Sheringham	Sand catchfly	*Silene conica*
83	09-06-21	Beeston Bump, nr Sheringham	Greater knapweed	*Centaurea scabiosa*
84	09-06-21	Kelling Heath, nr Cromer, Norfolk	Viper's bugloss	*Echium vulgare*
85	09-06-21	Kelling Heath, nr Cromer, Norfolk	Lesser stitchwort	*Stellaria graminea*
86	09-06-21	Edgefield Village Pond, S of Holt, Norfolk	Greater spearwort	*Ranunculus lingua*
87	09-06-21	Edgefield Village Pond, S of Holt, Norfolk	New Zealand pigmyweed	*Crassula helmsii*
88	09-06-21	Edgefield Village Pond, S of Holt, Norfolk	Water soldier	*Stratiotes aloides*
89	09-06-21	Edgefield Village Pond, S of Holt, Norfolk	Fringed water-lily	*Nymphoides peltatus*
90	09-06-21	Edgefield Village Pond, S of Holt, Norfolk	White water-lily	*Nymphaea alba*
91	09-06-21	Roper Farm, SE of Briston, Norfolk	Narrow-fruited cornsalad	*Valerianella dentata*
92	09-06-21	Roper Farm, SE of Briston, Norfolk	Corn marigold	*Chrysanthemum segetum*
93	09-06-21	Roper Farm, SE of Briston, Norfolk	Phacelia	*Phacelia tanacetafolia*
94	09-06-21	Roper Farm, SE of Briston, Norfolk	Field scabious	*Knautia arvensis*
95	09-06-21	Roper Farm, SE of Briston, Norfolk	Prickly poppy	*Papaver hybridum*
96	10-06-21	Base-rich flush, Widdybank Fell, Teesdale	Deer grass	*Trichophorum germanicum*
97	10-06-21	Base-rich flush, Widdybank Fell, Teesdale	Common butterwort	*Pinguicula vulgaris*
98	10-06-21	Base-rich flush, Widdybank Fell, Teesdale	Scottish asphodel	*Tofieldia pusilla*
99	10-06-21	Base-rich flush, Widdybank Fell, Teesdale	Bird's-eye primrose	*Primula farinosa*
00	10-06-21	Base-rich flush, Widdybank Fell, Teesdale	Teesdale sandwort	*Sabulina stricta*
01	10-06-21	Base-rich flush, Widdybank Fell, Teesdale	Dwarf milkwort	*Polygala amarella*
02	10-06-21	Exclosure, nr road, Widdybank Fell, Teesdale	Mountain pansy	*Viola lutea*
03	10-06-21	Exclosure, nr road, Widdybank Fell, Teesdale	Mountain everlasting	*Antennaria dioica*
04	10-06-21	Hay meadow next to Widdybank Fm	Zigzag clover	*Trifolium medium*
05	10-06-21	Hay meadow next to Widdybank Fm	Globeflower	*Trollius europaeus*
06	10-06-21	Hay meadow next to Widdybank Fm	Marsh valerian	*Valeriana dioica*
07	10-06-21	Hay meadow next to Widdybank Fm	Marsh lousewort	*Pedicularis palustris*
08	10-06-21	Hay meadow next to Widdybank Fm	Early marsh orchid	*Dactylorhiza incarnata*
09	10-06-21	Hay meadow next to Widdybank Fm	Broad-leaved cotton-grass	*Eriophorum latifolium*
10	10-06-21	Hay meadow next to Widdybank Fm	Alpine bartsia	*Bartsia alpina*
11	10-06-21	By roadside, Forest-in-Teesdale	Perennial cornflower	*Centaurea montana*
12	10-06-21	Cronkley Fell, Upper Teesdale NNR	Water avens	*Geum rivale*
13	10-06-21	Cronkley Fell, Upper Teesdale NNR	Juniper	*Juniperus communis*
14	10-06-21	Pile of Stones' Exclosure, Cronkley Fell	Hoary rockrose	*Helianthemum oelandicum*
15	10-06-21	Pile of Stones' Exclosure, Cronkley Fell	Horseshoe vetch	*Hippocrepis comosa*
16	10-06-21	Pile of Stones' Exclosure, Cronkley Fell	Blue-moor grass	*Sesleria caerulea*
17	10-06-21	Pile of Stones' Exclosure, Cronkley Fell	Purple moor-grass	*Molinia caerulea*

No.	Date	Location	Plant species - common name	Plant species - scientific name
418	10-06-21	Large exclosure by River Tees, Teesdale	Parsley fern	*Cryptogramma crispa*
419	10-06-21	Large exclosure by River Tees, Teesdale	Spring gentian	*Gentiana verna*
420	10-06-21	Large exclosure by River Tees, Teesdale	Oak fern	*Gymnocarpium dryopt*
421	10-06-21	Large exclosure by River Tees, Teesdale	Hard fern	*Blechnum spicant*
422	10-06-21	River bank, Cronkley Fell	Shrubby cinquefoil	*Potentilla fruticosa*
423	10-06-21	Stream joining the Tees, Cronkley Fell	Tawny sedge	*Carex hostiana*
424	10-06-21	Edge of Cronkley Pasture, Cronkley Fell	Yellow rattle	*Rhinanthus major*
425	10-06-21	Base-rich flush, Widdybank Fell, Teesdale	Marsh arrow-grass	*Triglochin palustris*
426	11-06-21	Base-rich flush, Widdybank Fell, Teesdale	Few-flowered spike-rush	*Eleocharis quinqueflora*
427	11-06-21	Road verge NW of Langdon Beck	Sweet cicely	*Myrrhis odorata*
428	11-06-21	Road verge NW of Langdon Beck	Wood cranesbill	*Geranium sylvaticum*
429	11-06-21	Road verge NW of Langdon Beck	Meadowsweet	*Filipendula ulmaria*
430	11-06-21	Uath Lochan Trail, Inshriach, Highland	Great wood-rush	*Luzula sylvatica*
431	11-06-21	Uath Lochan Trail, Inshriach, Highland	Chickweed wintergreen	*Trientalis europaea*
432	11-06-21	Uath Lochan Trail, Inshriach, Highland	Bog myrtle	*Myrica gale*
433	11-06-21	Uath Lochan Trail, Inshriach, Highland	Bogbean	*Menyanthes trifoliata*
434	11-06-21	Uath Lochan Trail, Inshriach, Highland	Cranberry	*Vaccinium oxycoccos*
435	12-06-21	Forest S of Carrbridge, Highland	Coralroot orchid	*Corallorhiza trifida*
436	12-06-21	Nr path, Dell of Abernethy, Highland	Twinflower	*Linnaea borealis*
437	12-06-21	Nr path, Dell of Abernethy, Highland	Common cow-wheat	*Melampyrum pratense*
438	12-06-21	SE of Boat of Garten, Abernethy NNR	Heath wood-rush	*Luzula multiflora*
439	12-06-21	SE of Boat of Garten, Abernethy NNR	Hairy wood-rush	*Luzula pilosella*
440	13-06-21	River Spey, GoS	Common valerian	*Valeriana officinalis*
441	13-06-21	River Spey, GoS	Nootka lupin	*Lupinus nootkatensis*
442	13-06-21	Old Military Road, Anagach Wood, GoS	Fox-and-cubs	*Pilosella aurantiaca*
443	13-06-21	Flowerfield, B970, N of Coylumbridge	Lesser butterfly orchid	*Platanthera bifolia*
444	13-06-21	Flowerfield, B970, N of Coylumbridge	Northern marsh orchid	*Dactylorhiza purpurell*
445	13-06-21	Flowerfield, B970, N of Coylumbridge	Heath fragrant orchid	*Gymnadenia borealis*
446	13-06-21	Behind visitor centre, Cairngorm car park	Cloudberry	*Rubus chamaemorus*
447	13-06-21	Path to N corries, Cairngorm	Dwarf cornel	*Cornus suecica*
448	14-06-21	SW end of Loch Duntelchaig, Highland	Star sedge	*Carex echinata*
449	14-06-21	SW end of Loch Duntelchaig, Highland	Oval sedge	*Carex leporina*
450	14-06-21	N bank of Beauly Firth, nr Red Castle	Frosted orache	*Atriplex laciniata*
451	14-06-21	N bank of Beauly Firth, nr Red Castle	Sea milkwort	*Glaux maritima*
452	15-06-21	Spey Bay, Moray Firth	Giant hogweed	*Heracleum mantegazzia*
453	15-06-21	E bank of River Spey, Spey Bay	Japanese knotweed	*Fallopia japonica*
454	15-06-21	Scrub at top of triangle, Spey Bay	Bittersweet	*Solanum dulcamara*
455	16-06-21	Walls of packhorse bridge, Carrbridge	Fairy foxglove	*Erinus alpinus*
456	16-06-21	Small dragonfly pond, nr Loch Garten RSPB	White sedge	*Carex canescens*
457	16-06-21	N side by beach, Chanonry Point	Marram grass	*Ammophila arenaria*
458	16-06-21	N side by beach, Chanonry Point	Lyme grass	*Leymus arenarius*
459	17-06-21	Stream, nr car park, Melon Udrigle	Marsh thistle	*Cirsium palustre*
460	18-06-21	Loch Flemington, nr Inverness	Amphibious bistort	*Persicaria amphibia*
461	18-06-21	Shingle, W of River Spey, Kingston, Moray	Weld	*Reseda luteola*
462	18-06-21	Shingle, W of River Spey, Kingston, Moray	Sticky groundsel	*Senecio viscosus*
463	18-06-21	Shingle, W of River Spey, Kingston, Moray	Welsh poppy	*Meconopsis cambrica*
464	18-06-21	Dune slack, W of River Spey, Kingston	Marsh bedstraw	*Galium palustre*
465	18-06-21	Dune slack, W of River Spey, Kingston	Marsh pennywort	*Hydrocotyle vulgaris*
466	18-06-21	Dune slack, W of River Spey, Kingston	Marsh cinquefoil	*Comarum palustre*
467	18-06-21	Dune slack, W of River Spey, Kingston	Creeping willow	*Salix repens*
468	18-06-21	Dune slack, W of River Spey, Kingston	Common cudweed	*Filago vulgaris*
469	19-06-21	Keltneyburn SWT reserve, nr Aberfeldy	Greater butterfly orchid	*Platanthera chlorantha*
470	19-06-21	Road verge, Keltneyburn SWT reserve	Marsh hawksbeard	*Crepis paludosa*
471	20-06-21	Clean Moor SSI, wood, NW of Taunton	Remote sedge	*Carex remota*

No.	Date	Location	Plant species - common name	Plant species - scientific name
2	19-06-21	Clean Moor SSI, wood, NW of Taunton	Field horsetail	*Equisetum arvense*
3	19-06-21	Clean Moor, SSI, mire, NW of Taunton	Hemp agrimony	*Eupatorium cannabinum*
4	19-06-21	Clean Moor, SSI, mire, NW of Taunton	Sharp-flowered rush	*Juncus acutiflorus*
5	19-06-21	Clean Moor, SSI, mire, NW of Taunton	Greater bird's-foot trefoil	*Lotus pedunculatus*
6	19-06-21	Clean Moor, SSI, mire, NW of Taunton	Parsley water dropwort	*Oenanthe lachenalii*
7	19-06-21	Clean Moor, SSI, mire, NW of Taunton	Bog pimpernel	*Anagellis tenella*
8	19-06-21	Clean Moor, SSI, mire, NW of Taunton	Blunt-flowered rush	*Juncus subnodulosus*
9	19-06-21	Clean Moor, SSI, mire, NW of Taunton	Marsh helleborine	*Epipactus palustris*
0	19-06-21	Clean Moor, SSI, mire, NW of Taunton	Fen bedstraw	*Galium uliginosum*
1	19-06-21	Clean Moor, SSI, mire, NW of Taunton	Meadow thistle	*Cirsium dissectum*
2	19-06-21	Clean Moor, SSI, mire, NW of Taunton	Carnation sedge	*Carex panicea*
3	19-06-21	Clean Moor, SSI, mire, NW of Taunton	Long-bracted yellow sedge	*Carex lepidocarpa*
4	19-06-21	Clean Moor, SSI, mire, NW of Taunton	Marsh fragrant orchid	*Gymnadenia densiflora*
5	19-06-21	Clean Moor, SSI, mire, NW of Taunton	Pale butterwort	*Pinguicula lusitanica*
6	19-06-21	Clean Moor, SSI, mire, NW of Taunton	Sneezewort	*Achillea ptarmica*
7	19-06-21	Clean Moor, SSI, mire, NW of Taunton	Tall fescue	*Schedornorus arundinaceus*
8	19-06-21	Clean Moor, SSI, mire, NW of Taunton	Hedge woundwort	*Stachys sylvatica*
9	19-06-21	Clean Moor SSI, mire edge, NW of Taunton	Lady fern	*Athyrium filix-femina*
0	19-06-21	Clean Moor SSI, mire edge, NW of Taunton	Common hemp-nettle	*Galeopsis tetrahit*
1	19-06-21	Clean Moor SSI, mire edge, NW of Taunton	Downy birch	*Betula pubescens*
2	19-06-21	Clean Moor SSI, mire edge, NW of Taunton	Bog stitchwort	*Stellaria alsine*
3	19-06-21	Clean Moor SSI, mire edge, NW of Taunton	Broad-leaved willowherb	*Epilobium montanum*
4	19-06-21	Clean Moor SSI, mire edge, NW of Taunton	Grey willow	*Salix cinerea* ssp. *oleifolia*
5	19-06-21	Clean Moor SSI, mire edge, NW of Taunton	Field rose	*Rosa arvensis*
6	19-06-21	Clean Moor SSI, mire edge, NW of Taunton	Golden-scaly male fern	*Dryopteris affinis*
7	19-06-21	Clean Moor SSI, mire edge, NW of Taunton	Blackcurrant	*Ribes nigrum*
8	19-06-21	Clean Moor SSI, mire edge, NW of Taunton	Wood dock	*Rumex sanguineus*
9	19-06-21	Clean Moor SSI, mire edge, NW of Taunton	Enchanter's nightshade	*Circaea lutetiana*
0	21-06-21	Walk to Fisherman's car park, Chew Stoke	Meadow cranesbill	*Geranium pratense*
1	21-06-21	Walk to Fisherman's car park, Chew Stoke	Meadow vetchling	*Lathyrus pratensis*
2	21-06-21	Walk to Fisherman's car park, Chew Stoke	False oat-grass	*Arrhenatherum elatius*
3	21-06-21	Walk to Fisherman's car park, Chew Stoke	Rough chervil	*Chaerophyllum temulum*
4	21-06-21	Walk to Fisherman's car park, Chew Stoke	Crested dogstail	*Cynosurus cristatus*
5	21-06-21	Walk to Fisherman's car park, Chew Stoke	Tufted vetch	*Vicia cracca*
6	21-06-21	Walk to Fisherman's car park, Chew Stoke	Spiked sedge	*Carex spicata*
7	21-06-21	Walk to Fisherman's car park, Chew Stoke	Meadow fescue	*Schedonorus pratensis*
8	22-06-21	Siston Brook, Cadbury Heath, Gloucs CC	Self-heal	*Prunella vulgaris*
9	22-06-21	Mangotsfield Cemetery, Gloucs	Rough hawkbit	*Leontodon hispidus*
0	22-06-21	Hester Wood Road, Yate, Gloucs	Field bindweed	*Convolvulus arvensis*
1	22-06-21	Road verge, Frogland Cross, Gloucs	Grass vetchling	*Lathyrus nissolia*
2	22-06-21	UWE and MOD roundabout, Bristol	Bee orchid	*Ophyris apifera*
3	22-06-21	UWE and MOD roundabout, Bristol	Pyramidal orchid	*Anacamptis pyramidalis*
4	22-06-21	UWE and MOD roundabout, Bristol	Smaller cat's-tail	*Phleum bertolonii*
5	23-06-21	Walk to Fisherman's car park, Chew Stoke	Corky-fruited water dropwort	*Oenanthe pimpinelloides*
6	23-06-21	Walk to Fisherman's car park, Chew Stoke	Nettle	*Urtica dioica*
7	23-06-21	Meadow nr Woodford Lodge, CV Lake	Betony	*Betonica officinalis*
8	23-06-21	Meadow nr Woodford Lodge, CV Lake	Devil's bit scabious	*Succisa pratensis*
9	23-06-21	Meadow nr Woodford Lodge, CV Lake	Common knapweed	*Centaurea nigra*
0	23-06-21	Meadow nr Fisherman's car park, CV Lake	Dyer's greenweed	*Genista tinctoria*
1	23-06-21	Westbury Road, Westbury-on-Trym, Bristol	Scented mayweed	*Matricaria chamomilla*
2	24-06-21	Knowle Hill, NE of CV Lake	Hairy St John's wort	*Hypericum hirsutum*
3	24-06-21	Knowle Hill, NE of CV Lake	Hedge bindweed	*Calystegia sepium*
4	24-06-21	Loop path, CV Lake	Creeping thistle	*Cirsium arvense*

No.	Date	Location	Plant species – common name	Plant species – scientific name
525	24-06-21	Loop path, CV Lake	Dog-rose	*Rosa canina*
526	24-06-21	Loop path, CV Lake	Marsh woundwort	*Stachys palustris*
527	24-06-21	Path between two car parks, CV Lake	Agrimony	*Agrimonia eupatoria*
528	24-06-21	Roadside, Wally Lane, by CV Lake	Bristly oxtongue	*Helminthotheca echio*
529	25-06-21	Yarley Fields Reserve, SWT, nr Bleadney	Yellow-wort	*Blackstonia perfoliata*
530	25-06-21	Yarley Fields Reserve, SWT, nr Bleadney	Quaking grass	*Briza media*
531	25-06-21	Yarley Fields Reserve, SWT, nr Bleadney	Yellow oat-grass	*Trisetum flavescens*
532	25-06-21	Yarley Fields Reserve, SWT, nr Bleadney	Restharrow	*Ononis repens*
533	25-06-21	Yarley Fields Reserve, SWT, nr Bleadney	Meadow barley	*Hordeum secalinum*
534	25-06-21	Yarley Fields Reserve, SWT, nr Bleadney	Common centaury	*Centaurium erythraea*
535	25-06-21	Yarley Fields Reserve, SWT, nr Bleadney	Wild privet	*Ligustrum vulgare*
536	25-06-21	Yarley Fields Reserve, SWT, nr Bleadney	Hoary plantain	*Plantago media*
537	25-06-21	Boggy area, Yarley Fields Reserve SWT	Hard rush	*Juncus inflexus*
538	25-06-21	Boggy area, Yarley Fields Reserve SWT	False fox sedge	*Carex otrubae*
539	26-06-21	W of Cannop Ponds, Forest of Dean, Gloucs	Water-pepper	*Persicaria hydropiper*
540	27-06-21	Friary to Freshford Mill, nr Freshford	Bath asparagus	*Ornthithogalum pyrenaicum*
541	27-06-21	Friary to Freshford Mill, nr Freshford	Large bindweed	*Calystegia sylvatica*
542	27-06-21	The Tyning, Freshford, nr Bath	Crow garlic	*Allium vineale*
543	27-06-21	The Tyning, Freshford, nr Bath	Rosebay willowherb	*Chamerion angustifoli*
544	27-06-21	The Tyning, Freshford, nr Bath	Bearded couch	*Elymus caninus*
545	27-06-21	Freshford Village, nr Bath	Love-in-a-mist	*Nigella damascena*
546	27-06-21	Freshford Village, nr Bath	Yellow loosestrife	*Lysimachia vulgaris*
547	27-06-21	Freshford Village, nr Bath	Opium poppy	*Papaver somniferum*
548	28-06-21	Tom's Farm, nr Linwood, NF	Mousetail	*Myosurus minimus*
549	28-06-21	Tom's Farm, nr Linwood, NF	Common knotgrass	*Polygonum aviculare*
550	28-06-21	Verge, Broomy Plain, road to Lyndhurst	Trailing St John's wort	*Hypericum humifusu*
551	28-06-21	Verge, Broomy Plain, road to Lyndhurst	Cross-leaved heath	*Erica tetralix*
552	28-06-21	Verge, Broomy Plain, road to Lyndhurst	Lesser hawkbit	*Leontodon saxatilis*
553	28-06-21	Seasonal pond, nr Bolton's Bench, Lyndhurst	Chamomile	*Chamaemelum nobile*
554	28-06-21	Seasonal pond, nr Bolton's Bench, Lyndhurst	Slender marsh bedstraw	*Galium constrictum*
555	28-06-21	Seasonal pond, nr Bolton's Bench, Lyndhurst	Marsh cudweed	*Gnaphalium uliginos*
556	28-06-21	White Moor, by B3056, E of Lyndhurst	Marsh St John's wort	*Hypericum elodes*
557	28-06-21	White Moor, by B3056, E of Lyndhurst	Oblong-leaved sundew	*Drosera intermedia*
558	28-06-21	White Moor, by B3056, E of Lyndhurst	Lesser skullcap	*Scutellaria minor*
559	28-06-21	Matley Bog, W of Beaulieu Road Station	Bog asphodel	*Narthecium ossifragu*
560	28-06-21	Matley Bog, W of Beaulieu Road Station	Bog orchid	*Hammarbya paludosa*
561	28-06-21	Road between Exbury Gdns and Beaulieu Heath	Narrow-leaved lungwort	*Pulmonaria longifolia*
562	28-06-21	East End Ponds, NF	Pennyroyal	*Mentha pulegium*
563	28-06-21	East End Ponds, NF	Redshank	*Persicaria maculosa*
564	28-06-21	East End Ponds, NF	Hampshire purslane	*Ludwigia palustris*
565	28-06-21	East End Ponds, NF	Coral necklace	*Illecebrum vertillicatu*
566	28-06-21	East End Ponds, NF	Four-leaved allseed	*Polycarpon tetraphyllu*
567	28-06-21	Crockford Bridge, Beaulieu Heath	Yellow centaury	*Cicendia filiformis*
568	28-06-21	Pond, Crockford Bridge, Beaulieu Heath	Creeping forget-me-not	*Myosotis secunda*
569	28-06-21	Roydon Woods NR, S of Brockenhurst	Bastard balm	*Melittis melissophyllu*
570	28-06-21	Roydon Woods NR, S of Brockenhurst	Compact rush	*Juncus conglomeratus*
571	28-06-21	Stony Moors, nr Holmsley Inclosure, NF	Marsh willowherb	*Epilobium palustre*
572	28-06-21	Stony Moors, nr Holmsley Inclosure, NF	Great sundew	*Drosera anglica*
573	28-06-21	Burley New Enclosure, NF	Wild gladiolus	*Gladiolus illyricus*
574	29-06-21	Cemlyn Bay, NWWT Reserve, Anglesey	Sea-kale	*Crambe maritima*
575	29-06-21	Road verge, Talwrn, Anglesey	Frog orchid	*Coeloglossum viride*
576	29-06-21	Wood by verge, Talwrn, Anglesey	Wilson's honeysuckle	*Lonicera nitida*
577	30-06-21	Ty Mawr, nr South Stack, Anglesey	Great mullein	*Verbascum thapsus*
578	30-06-21	Car park, South Stack RSPB, Angelsey	Hemlock	*Conium maculatum*

o.	Date	Location	Plant species - common name	Plant species - scientific name
9	30-06-21	South Stack RSPB, Holy Island, Anglesey	Silver hair-grass	*Aira caryophyllea*
0	30-06-21	South Stack RSPB, Holy Island, Anglesey	Burnet-saxifrage	*Pimpinella saxifraga*
1	30-06-21	South Stack RSPB, Holy Island, Anglesey	Western gorse	*Yulex gallii*
2	30-06-21	South Stack RSPB, Holy Island, Anglesey	Green-ribbed sedge	*Carex binervis*
3	30-06-21	South Stack RSPB, Holy Island, Anglesey	Slender St John's wort	*Hypericum pulchrum*
4	30-06-21	South Stack RSPB, Holy Island, Anglesey	Seaside centaury	*Centaureum littorale*
5	30-06-21	South Stack RSPB, Holy Island, Anglesey	Heath grass	*Danthonia decumbens*
6	30-06-21	South Stack RSPB, Holy Island, Anglesey	Goldenrod	*Solidago virgaurea*
7	30-06-21	South Stack RSPB, Holy Island, Anglesey	Hay-scented buckler fern	*Dryopteris aemula*
8	30-06-21	South Stack RSPB, Holy Island, Anglesey	Mat-grass	*Nardus stricta*
9	30-06-21	South Stack RSPB, Holy Island, Anglesey	Spotted rockrose	*Tubaria guttata*
0	30-06-21	South Stack RSPB, Holy Island, Anglesey	Wild angelica	*Angelica sylvestris*
1	30-06-21	South Stack RSPB, Holy Island, Anglesey	Golden samphire	*Inula crithmoides*
2	30-06-21	South Stack RSPB, Holy Island, Anglesey	Portland spurge	*Euphorbia portlandica*
3	30-06-21	South Stack RSPB, Holy Island, Anglesey	Fern grass	*Catapodium rigidum*
4	30-06-21	South Stack RSPB, Holy Island, Anglesey	Field (spathulate) fleawort	*Tephroseris integrifolia* ssp. *maritimus*
5	30-06-21	Aberffraw dunes, nr Aberffraw, Anglesey	Goat's-beard	*Tragopogon pratensis*
6	30-06-21	Aberffraw dunes, nr Aberffraw, Anglesey	Marsh horsetail	*Equisetum palustre*
7	30-06-21	Aberffraw dunes, nr Aberffraw, Anglesey	Common fleabane	*Pulicaria dysenterica*
8	30-06-21	Aberffraw dunes, nr Aberffraw, Anglesey	Strawberry clover	*Trifolium fragiferum*
9	30-06-21	Aberffraw dunes, nr Aberffraw, Anglesey	Corn sow-thistle	*Sonchus arvensis*
0	30-06-21	Aberffraw dunes, nr Aberffraw, Anglesey	Common bent	*Agrostis capillaris*
1	30-06-21	Aberffraw dunes, nr Aberffraw, Anglesey	Houndstongue	*Cynoglossum officinale*
2	30-06-21	Aberffraw dunes, nr Aberffraw, Anglesey	Blue fleabane	*Erigeron acris*
3	30-06-21	Aberffraw dunes, nr Aberffraw, Anglesey	Wild pansy	*Viola tricolor* ssp. *curtisii*
4	30-06-21	Aberffraw dunes, nr Aberffraw, Anglesey	Squirrel-tail fescue	*Vulpia bromoides*
5	30-06-21	Aberffraw dunes, nr Aberffraw, Anglesey	Sand cat's-tail	*Phleum arenarium*
6	30-06-21	Aberffraw dunes, nr Aberffraw, Anglesey	Variegated horsetail	*Equisetum variegatum*
7	30-06-21	Aberffraw dunes, nr Aberffraw, Anglesey	Lesser clubmoss	*Selaginella selaginoides*
8	30-06-21	Aberffraw dunes, nr Aberffraw, Anglesey	Round-leaved wintergreen	*Pyrola rotundifolia*
9	30-06-21	Aberffraw dunes, nr Aberffraw, Anglesey	Early sand-grass	*Mibora minima*
0	30-06-21	Aberffraw dunes, nr Aberffraw, Anglesey	Burnet rose	*Rosa spinosissima*
1	30-06-21	Aberffraw dunes, nr Aberffraw, Anglesey	Autumn gentian	*Gentianella amarella*
2	30-06-21	Aberffraw dunes, nr Aberffraw, Anglesey	Water-cress	*Nasturtium officinale*
3	30-06-21	Llanddwyn Island, Newborough Warren	Horseradish	*Armoracia rusticana*
4	30-06-21	Llanddwyn Island, Newborough Warren	Maiden pink	*Dianthus deltoides*
5	30-06-21	Llanddwyn Island, Newborough Warren	Ivy broomrape	*Orobanche hederae*
6	30-06-21	Llanddwyn Island, Newborough Warren	Lanceolate spleenwort	*Asplenium obovatum*
7	30-06-21	Llanddwyn Island, Newborough Warren	Stinking tutsan	*Hypericum hircinum*
8	30-06-21	Llanddwyn Island, Newborough Warren	Small-flowered evening primrose	*Oenothera cambrica*
9	30-06-21	Llanddwyn Island, Newborough Warren	Duke of Argyll's teaplant	*Lycium barbarum*
0	30-06-21	Newborough Forest, Newborough Warren	Dune helleborine	*Epipactus dunensis*
1	30-06-21	Newborough Forest, Newborough Warren	Narrow-leaved helleborine	*Cephalanthera longifolia*
2	01-07-21	Cors Goch NWWT Reserve, Anglesey	Common gromwell	*Lithospermum officinale*
3	01-07-21	Cors Goch NWWT Reserve, Anglesey	Common figwort	*Scrophularia nodosa*
4	01-07-21	Cors Goch NWWT Reserve, Anglesey	Great fen-sedge	*Cladium mariscus*
5	01-07-21	Cors Goch NWWT Reserve, Anglesey	Slender sedge	*Carex lasiocarpa*
6	01-07-21	Cors Goch NWWT Reserve, Anglesey	Lesser bladderwort	*Utricularia minor*
7	01-07-21	Cors Goch NWWT Reserve, Anglesey	Narrow buckler-fern	*Dryopteris carthusiana*
8	01-07-21	Cors Goch NWWT Reserve, Anglesey	Bottle sedge	*Carex rostrata*
9	01-07-21	Ty'n-Y-Coed Road, Great Orme	Musk storksbill	*Erodium moschatum*
0	01-07-21	Ty'n-Y-Coed Road, Great Orme	Knotted hedge-parsley	*Torilis nodosa*

No.	Date	Location	Plant species - common name	Plant species - scientific name
631	01-07-21	Limestone grassland, S of mine, Great Orme	Black horehound	*Ballota nigra*
632	01-07-21	Limestone grassland, S of mine, Great Orme	Spearmint	*Mentha spicata*
633	01-07-21	Limestone grassland, S of mine, Great Orme	Downy oat-grass	*Avenula pubescens*
634	01-07-21	Limestone grassland, S of mine, Great Orme	Meadow oat-grass	*Avenula pratensis*
635	01-07-21	Limestone grassland, S of mine, Great Orme	Confused hawkweed	*Hieracium britannicoid*
636	01-07-21	Limestone grassland, S of mine, Great Orme	Spiked speedwell	*Veronica spicata*
637	01-07-21	Limestone grassland, S of mine, Great Orme	Moonwort	*Botrychium lunaria*
638	01-07-21	Limestone grassland, S of mine, Great Orme	Wild cotoneaster	*Cotoneaster cambricus*
639	01-07-21	Limestone grassland, S of mine, Great Orme	Small scabious	*Scabiosa columbaria*
640	01-07-21	Limestone grassland, S of mine, Great Orme	Carline thistle	*Carlina vulgaris*
641	01-07-21	Limestone grassland, S of mine, Great Orme	White horehound	*Marrubium vulgare*
642	01-07-21	Ty'n-Y-Coed Road, Great Orme	Feverfew	*Tanacetum parthenium*
643	01-07-21	Ty'n-Y-Coed Road, Great Orme	Purple toadflax	*Linaria purpurea*
644	01-07-21	Ty'n-Y-Coed Road, Great Orme	Himalayan giant bramble	*Rubus armeniacus*
645	01-07-21	Viewpoint nr Haulfre Gardens, Great Orme	Rock whitebeam	*Sorbus rupicola*
646	01-07-21	Viewpoint nr Haulfre Gardens, Great Orme	Slender thistle	*Carduus tenuiflorus*
647	01-07-21	Viewpoint nr Haulfre Gardens, Great Orme	Ploughman's spikenard	*Inula conyzae*
648	01-07-21	Viewpoint nr Haulfre Gardens, Great Orme	Hoary mustard	*Hirschfeldia incana*
649	01-07-21	Viewpoint nr Haulfre Gardens, Great Orme	Dark red helleborine	*Epipactus atrorubens*
650	01-07-21	Viewpoint nr Haulfre Gardens, Great Orme	Purple-flushed hawkweed	*Hieracium pseudolayi*
651	01-07-21	Viewpoint nr Haulfre Gardens, Great Orme	Marjoram	*Origanum vulgare*
652	01-07-21	Viewpoint nr Haulfre Gardens, Great Orme	Small-leaved sweet-briar	*Rosa agrestis*
653	01-07-21	Viewpoint nr Haulfre Gardens, Great Orme	Sweet-briar	*Rosa rubiginosa*
654	01-07-21	Viewpoint nr Haulfre Gardens, Great Orme	Rock cotoneaster	*Cotoneaster integrifoliu*
655	01-07-21	Viewpoint nr Haulfre Gardens, Great Orme	Harebell	*Campanula rotundifoli*
656	01-07-21	Viewpoint nr Haulfre Gardens, Great Orme	Nottingham catchfly	*Silene nutans*
657	02-07-21	The Hermitage, nr Dunkeld, Perthshire	Ground elder	*Aegopodium podagrari*
658	04-07-21	Saltmarsh, Findhorn Bay LNR, Moray	Sea aster	*Aster tripolium*
659	04-07-21	Saltmarsh, Findhorn Bay LNR, Moray	Greater sea-spurrey	*Spergularia media*
660	04-07-21	Saltmarsh, Findhorn Bay LNR, Moray	Scot's lovage	*Ligusticum scoticum*
661	04-07-21	Saltmarsh, Findhorn Bay LNR, Moray	Red fescue	*Restuca rubra*
662	05-07-21	Small burn, Roseisle Forest, Moray	Branched bur-reed	*Sparganium erectum*
663	06-07-21	Golf course path, Anagach Wood, GoS	Intermediate wintergreen	*Pyrola media*
664	06-07-21	Beyond feeders, Anagach Wood, GoS	Creeping lady's-tresses	*Goodyera repens*
665	06-07-21	Nr Pine marten hide, B970, Inshriach House	Jacob's-ladder	*Polemonium caeruleum*
666	06-07-21	Nr Pine marten hide, B970, Inshriach House	Monkey flower	*Erythranthe guttata*
667	07-07-21	Path to Coire an t-Sneachda, Cairngorm	Lesser twayblade	*Neottia cordata*
668	07-07-21	Path to Coire an t-Sneachda, Cairngorm	Heath rush	*Juncus squarrosus*
669	07-07-21	Path to Coire an t-Sneachda, Cairngorm	Fir clubmoss	*Huperzia selago*
670	07-07-21	Path to Coire an t-Sneachda, Cairngorm	Alpine lady's mantle	*Alchemilla alpina*
671	07-07-21	Path to Coire an t-Sneachda, Cairngorm	Bearberry	*Arctostaphylos uva-ursi*
672	07-07-21	Wet flush, Coire an t-Sneachda, Cairngorm	Starry saxifrage	*Saxifraga stellaris*
673	07-07-21	Above 750m, Coire an t-Sneachda	Alpine clubmoss	*Diphasiastrum alpinum*
674	07-07-21	Above 750m, Coire an t-Sneachda	Bog bilberry	*Vaccinium uliginosum*
675	07-07-21	Above 900m, Coire an t-Sneachda	Roseroot	*Sedum rosea*
676	07-07-21	Above 900m, Coire an t-Sneachda	Dwarf cudweed	*Omalotheca supina*
677	07-07-21	Above 900m, Coire an t-Sneachda	Trailing azalea	*Kalmia procumbens*
678	07-07-21	Above 900m, Coire an t-Sneachda	Marsh violet	*Viola palustris*
679	07-07-21	Above 900m, Coire an t-Sneachda	Common mouse-ear	*Cerastium fontanum*
680	07-07-21	Wet flush, Coire an t-Sneachda, Cairngorm	Alpine bistort	*Persicaria vivipara*
681	07-07-21	Above 800m, Coire an t-Sneachda	Three-leaved rush	*Juncus trifidus*
682	07-07-21	Above 750m, Coire an t-Sneachda	Small cow-wheat	*Melampyrum sylvaticu*
683	07-07-21	Above 750m, Coire an t-Sneachda	Viviparous fescue	*Festuca vivipara*
684	08-07-21	By River Spey, FP between bridges, GoS	Melancholy thistle	*Cirsium heterophyllum*

o.	Date	Location	Plant species - common name	Plant species - scientific name
5	08-07-21	By River Spey, FP between bridges, GoS	Northern bedstraw	*Galium boreale*
6	08-07-21	By River Spey, FP between bridges, GoS	Perforate St John's wort	*Hypericum perforatum*
7	08-07-21	Top car park, Coignafearn, Findhorn Valley	Small white orchid	*Pseudorchis albida*
8	10-07-21	Saltmarsh, Loch Fleet NNR, nr Golspie	Common glasswort	*Salicornia europaea*
9	10-07-21	Saltmarsh, Loch Fleet NNR, nr Golspie	Saltmarsh rush	*Juncus gerardii*
0	10-07-21	Saltmarsh, Loch Fleet NNR, nr Golspie	Red bartsia	*Odontites verna*
1	10-07-21	Balblair Wood, nr Golspie	One-flowered wintergreen	*Moneses uniflora*
2	11-07-21	Badgeworth SSSI, GWT Reserve, Gloucs	Adder's-tongue spearwort	*Ranunculus ophioglossifolius*
3	11-07-21	Badgeworth SSSI, GWT Reserve, Gloucs	Narrow-leaved water-plantain	*Alisma lanceolatum*
4	11-07-21	Badgeworth SSSI, GWT Reserve, Gloucs	Water speedwell	*Veronica anagallis-aquatica*
5	11-07-21	Badgeworth SSSI, GWT Reserve, Gloucs	Creeping-Jenny	*Lysimachia nummularia*
6	11-07-21	Badgeworth SSSI, GWT Reserve, Gloucs	Common water-crowfoot	*Ranunculus aquatilis*
7	11-07-21	Badgeworth SSSI, GWT Reserve, Gloucs	Creeping bent	*Agrostis stolonifera*
8	11-07-21	Badgeworth SSSI, GWT Reserve, Gloucs	Pink water-speedwell	*Veronica catenata*
9	11-07-21	Old London Road, GWT Reserve, Gloucs	Limestone woundwort	*Stachys alpina*
0	11-07-21	Old London Road, GWT Reserve, Gloucs	Nettle-leaved bellflower	*Campanula trachelium*
1	11-07-21	B4509, Tortworth, J14 of M5	Common blue-sowthistle	*Cicerbita macrophylla*
2	13-07-21	Walk to Fisherman's car park, Chew Stoke	False-brome	*Brachypodium sylvaticum*
3	13-07-21	Knowle Hill, NE of CV Lake	Welted thistle	*Carduus crispus*
4	13-07-21	Path between two car parks, CV Lake	Dewberry	*Rubus caesius*
5	13-07-21	Gully, by Circular road, Clifton Downs	Basil thyme	*Clinopodium acinos*
6	13-07-21	Gully, by Circular road, Clifton Downs	Round-headed leek	*Allium sphaerocephalon*
7	13-07-21	Gully, by Circular road, Clifton Downs	Musk thistle	*Carduus nutans*
8	14-07-21	Calstone and Cherhill Downs NT, Wilts	Wooly thistle	*Cirsium eriophorum*
9	14-07-21	Calstone and Cherhill Downs NT, Wilts	Dwarf thistle	*Cirsium acaule*
0	14-07-21	Calstone and Cherhill Downs NT, Wilts	Squinancywort	*Asperula cynanchica*
1	14-07-21	Calstone and Cherhill Downs NT, Wilts	Round-headed rampion	*Phyteuma orbiculare*
2	14-07-21	Calstone and Cherhill Downs NT, Wilts	Tor-grass	*Brachypodium rupestre*
3	14-07-21	Whitehorse Trail, C and C Downs, Wilts	Large thyme	*Thymus pulegioides*
4	14-07-21	Whitehorse Trail, C and C Downs, Wilts	Chalk fragrant-orchid	*Gymnadenia conopsea*
5	14-07-21	Roundabout A338/M4, N of Hungerford	Broad-leaved everlasting pea	*Lathyrus latifolius*
6	14-07-21	Roundabout A338/M4, N of Hungerford	Goat's rue	*Galega officinalis*
7	14-07-21	Roundabout A338/M4, N of Hungerford	Wild parsnip	*Pastinaca sativa*
8	14-07-21	Roundabout A338/M4, N of Hungerford	Mugwort	*Artemisia vulgaris*
9	14-07-21	Roundabout A338/M4, N of Hungerford	Ribbed melilot	*Melilotus officinalis*
0	14-07-21	M4 E, nr Reading	Hollyhock	*Alcea rosea*
1	15-07-21	Entrance, Ranscombe Farm PL Reserve	Hedge parsley	*Torilis japonica*
2	15-07-21	Entrance, Ranscombe Farm PL Reserve	Wild basil	*Clinipodium vulgare*
3	15-07-21	Arable fields, Ranscombe Farm PL Reserve	Stinking chamomile	*Anthemis cotula*
4	15-07-21	Arable, N Downs, Ranscombe Farm	Wild oat	*Avena fatua*
5	15-07-21	Arable, N Downs, Ranscombe Farm	Broad-leaved cudweed	*Filago pyramidata*
6	15-07-21	Arable, N Downs, Ranscombe Farm	Vervain	*Verbena officinalis*
7	15-07-21	Arable, N Downs, Ranscombe Farm	Sun spurge	*Euphorbia helioscopia*
8	15-07-21	Arable, N Downs, Ranscombe Farm	Interrupted brome	*Bromus interruptus*
9	15-07-21	Arable, N Downs, Ranscombe Farm	Henbit dead-nettle	*Lamium amplexicaule*
0	15-07-21	Arable, N Downs, Ranscombe Farm	Rye brome	*Bromus secalinus*
1	15-07-21	Arable, N Downs, Ranscombe Farm	Great brome	*Anisantha diandra*
2	15-07-21	Arable, N Downs, Ranscombe Farm	Pheasant's eye	*Adonis annua*
3	15-07-21	Arable, N Downs, Ranscombe Farm	Dwarf spurge	*Euphorbia exigua*
4	15-07-21	Arable, Kitchen Field, Ranscombe Farm	Bastard cabbage	*Rapistrum rogosum*
5	15-07-21	Arable, Kitchen Field, Ranscombe Farm	Meadow clary	*Salvia pratensis*

No.	Date	Location	Plant species - common name	Plant species - scientific name
736	15-07-21	Arable, Kitchen Field, Ranscombe Farm	Wild liquorice	*Astragalus glycyphyllo*
737	15-07-21	Arable, Kitchen Field, Ranscombe Farm	Man orchid	*Orchis anthropophora*
738	15-07-21	Arable, Kitchen Field, Ranscombe Farm	Ground-pine	*Ajuga chamaepitys*
739	15-07-21	Arable, Kitchen Field, Ranscombe Farm	Blue pimpernel	*Lysimachia foemina*
740	15-07-21	Arable, Kitchen Field, Ranscombe Farm	Rough mallow	*Malva setigera*
741	15-07-21	Arable, Kitchen Field, Ranscombe Farm	Small-flowered sweet-briar	*Rosa micrantha*
742	15-07-21	Arable, Kitchen Field, Ranscombe Farm	Trailing tormentil	*Potentilla anglica*
743	15-07-21	Arable, Kitchen Field, Ranscombe Farm	Corn mint	*Mentha arvensis*
744	15-07-21	Arable, Kitchen Field, Ranscombe Farm	Hawkweed oxtongue	*Picris hieracioides*
745	15-07-21	Arable, Kitchen Field, Ranscombe Farm	Cultivated flax	*Linum usitatissimum*
746	15-07-21	Arable, The Valley, Ranscombe Farm	Corncockle	*Agrostemma githago*
747	15-07-21	Arable, The Valley, Ranscombe Farm	Stone parsley	*Sison amomum*
748	15-07-21	Arable, The Valley, Ranscombe Farm	Venus's looking-glass	*Legousia hybrida*
749	15-07-21	Arable, The Valley, Ranscombe Farm	Common fumitory	*Fumaria officinalis*
750	15-07-21	Arable, The Valley, Ranscombe Farm	Dense-flowered fumitory	*Fumaria densiflora*
751	15-07-21	Arable, The Valley, Ranscombe Farm	Rough poppy	*Roemeria hispida*
752	15-07-21	Arable, The Valley, Ranscombe Farm	Black bindweed	*Fallopia convolvulus*
753	15-07-21	Chalk grassland, Ranscombe Farm	Deadly nightshade	*Atropa belladonna*
754	15-07-21	Chalk grassland, Ranscombe Farm	Clustered bellflower	*Campanula glomerat*
755	15-07-21	Chalk grassland, Ranscombe Farm	Hoary ragwort	*Jacobaea erucifolia*
756	15-07-21	Woodland edge, Ranscombe Farm	Broad-leaved helleborine	*Epipactus helleborine*
757	15-07-21	Boxley Wood, N of Boxley, M20	Stinking iris	*Iris foetidissima*
758	15-07-21	Boxley Wood, N of Boxley, M20	Yellow bird's-nest	*Hypopitys monotropa*
759	15-07-21	Road verge, A2, NW of Dover	Chicory	*Cichorium intybus*
760	16-07-21	Restharrow Dunes, nr Sandwich Bay, Kent	Lizard orchid	*Himantoglossum hirci*
761	16-07-21	Restharrow Dunes, nr Sandwich Bay, Kent	Fennel	*Foeniculum vulgare*
762	16-07-21	Betteshanger Park, nr Deal	Smooth tare	*Ervum tetrasperma*
763	16-07-21	Betteshanger Park, nr Deal	Wall bedstraw	*Galium parisiense*
764	16-07-21	Betteshanger Park, nr Deal	Lesser bulrush	*Typha angustifolia*
765	16-07-21	Betteshanger Park, nr Deal	Moth mullein	*Verbascum blattaria*
766	16-07-21	Along wall, Beach Road, Deal	Round-leaved fluellen	*Kickxia spuria*
767	16-07-21	Samphire Hoe NR, SW of Dover	Rock sea-lavender	*Limonium binervosum*
768	16-07-21	South Foreland, S of St Margaret's Bay	Rose-of-Sharon	*Hypericum calycinum*
769	16-07-21	South Foreland, S of St Margaret's Bay	Chalk milkwort	*Polygala calcarea*
770	16-07-21	South Foreland, S of St Margaret's Bay	Common toadflax	*Linaria vulgaris*
771	16-07-21	South Foreland, S of St Margaret's Bay	Common broomrape	*Orobanche minor*
772	16-07-21	Beach, St Margaret's Bay	Russian-vine	*Fallopia baldschuanica*
773	16-07-21	The Old Rifle Range, S of Deal	Narrow-leaved bird's-foot trefoil	*Lotus tenuis*
774	16-07-21	The Old Rifle Range, S of Deal	Oxtongue broomrape	*Orobanche picridis*
775	18-07-21	Monkswood NNR, Hunts	Crested cow-wheat	*Melampyrum cristatum*
776	18-07-21	Monkswood NNR, Hunts	Small teasel	*Dipsacus pilosus*
777	19-07-21	East Harptree Woods, nr East Harptree	Square-stalked St John's wort	*Hypericum tetrapterum*
778	21-07-21	Climb up ridge, Cùl Mòr, N of Ullapool	Dwarf willow	*Salix herbacea*
779	21-07-21	Plateau, NW of Cùl Mòr summit	Norwegian mugwort	*Artemisia norvegica*
780	21-07-21	Plateau, NW of Cùl Mòr summit	Cyphel	*Minuartia sedoides*
781	21-07-21	On descent, E of Cùl Mòr summit	Beech fern	*Phegopteris connectilis*
782	22-07-21	Culvert, by road, N of Cùl Mòr	Yellow saxifrage	*Saxifraga aizoides*
783	22-07-21	Culvert, by road, N of Cùl Mòr	New Zealand willowherb	*Epilobium brunnescen*
784	22-07-21	Culvert, by road, N of Cùl Mòr	Holly fern	*Polystichum lonchitis*
785	22-07-21	Culvert, by road, N of Cùl Mòr	Green spleenwort	*Asplenium viride*
786	22-07-21	Culvert, by road, N of Cùl Mòr	Brittle bladder-fern	*Cystopteris fragilis*
787	22-07-21	A837, W of Oykel Bridge, nr Lairg	Hare's-tail cottongrass	*Eriophorum vaginatum*
788	22-07-21	A837, E of Oykel Bridge, nr Lairg	Confused bridewort	*Spiraea x pseudosalici*

o.	Date	Location	Plant species - common name	Plant species - scientific name
89	22-07-21	Shingle beach, N of Murkle, nr Thurso	Oysterplant	*Mertensia maritima*
90	22-07-21	Shingle beach, N of Murkle, nr Thurso	Common orache	*Atriple patula*
91	22-07-21	Coronation Meadow, nr Dunnet Head	Grass-of-Parnassus	*Parnassia palustris*
92	22-07-21	Coronation Meadow, nr Dunnet Head	Scottish primrose	*Primula scotica*
93	22-07-21	Broubster Leans, RSPB Reserve, nr Thurso	Marsh ragwort	*Jacobaea aquatica*
94	22-07-21	Broubster Leans, RSPB Reserve, nr Thurso	Narrow small-reed	*Calamagrostis stricta*
95	22-07-21	Park Terrace, Brora, Caithness	Northern Dock	*Rumex longifolius*
96	22-07-21	Park Terrace, Brora, Caithness	Long-headed poppy	*Papaver dubium*
97	23-07-21	Ridge-top plateau, Cairn Gorm	Moss campion	*Silene acaulis*
98	23-07-21	Ridge-top plateau, Cairn Gorm	Northern rock-cress	*Arabidopsis petraea*
99	23-07-21	Arable meadow, nr Nethy Bridge	Cornflower	*Centaurea cyanus*
00	23-07-21	W of A939, Anagach Wood, GoS	Pink purslane	*Claytonia sibirica*
01	24-07-21	Field behind Glenshee Ski Centre, A93	Knotted pearlwort	*Sagina nodosa*
02	24-07-21	Field behind Glenshee Ski Centre, A93	Stagshorn clubmoss	*Lycopodium clavatum*
03	24-07-21	Field behind Glenshee Ski Centre, A93	Norwegian moonwort	*Botrychium nordicum*
04	24-07-21	Wet flushes on hill, Glenshee	Eared willow	*Salix aurita*
05	24-07-21	Wet flushes on hill, Glenshee	Alpine meadow-rue	*Thalictrum alpinum*
06	24-07-21	Wet flushes on hill, Glenshee	Three-flowered rush	*Juncus triglumis*
07	24-07-21	Wet flushes on hill, Glenshee	Chickweed willowherb	*Epilobium alsinifolium*
08	24-07-21	Wet flushes on hill, Glenshee	Mountain sibbaldia	*Sibbaldia procumbens*
09	24-07-21	Wet flushes on hill, Glenshee	Lemon-scented fern	*Oreopteris limbosperma*
10	24-07-21	Base-rich flush at high altitude, Glenshee	Hair sedge	*Carex capillaris*
11	24-07-21	Base-rich flush at high altitude, Glenshee	Dioecious sedge	*Carex dioica*
12	24-07-21	Sugar limestone, Glenshee	Mountain avens	*Dryas octopetala*
13	24-07-21	Sugar limestone, Glenshee	Rock sedge	*Carex rupestris*
14	24-07-21	Next to sugar limestone, Glenshee	Sheathed sedge	*Carex vaginata*
15	24-07-21	Next to sugar limestone, Glenshee	Alpine saw-wort	*Saussurea alpina*
16	24-07-21	Next to sugar limestone, Glenshee	Mountain sandwort	*Minuartia rubella*
17	24-07-21	Next to sugar limestone, Glenshee	Hoary whitlowgrass	*Draba incana*
18	24-07-21	Grassland saddle below summit, Glenshee	Alpine milk-vetch	*Astragalus alpinus*
19	24-07-21	In valley nr a burn, Glenshee	Alpine speedwell	*Veronica alpina*
20	24-07-21	Grassland saddle below summit, Glenshee	Shade horsetail	*Equisetum pratense*
21	25-07-21	Upland pasture, Ben Lawers NNR	Spignel	*Meum athamanticum*
22	25-07-21	Upland pasture, Ben Lawers NNR	Brown bent	*Agrostis vinealis*
23	25-07-21	Upland pasture, Ben Lawers NNR	Jointed rush	*Juncus articulatus*
24	25-07-21	Base-rich bog pool, Ben Lawers NNR	Water sedge	*Carex aquatilis*
25	25-07-21	Base-rich burn, Ben Lawers NNR	Stiff sedge	*Carex bigelowii*
26	25-07-21	Base-rich burn, Ben Lawers NNR	Sheep's fescue	*Festuca ovalis*
27	25-07-21	Base-rich burn, Ben Lawers NNR	Russet sedge	*Carex saxatilis*
28	25-07-21	Base-rich burn, Ben Lawers NNR	Alpine willowherb	*Epilobium anagallidifolium*
29	25-07-21	Base-rich burn, Ben Lawers NNR	Mountain scurvy-grass	*Cochlearia micacea*
30	25-07-21	Path to mica schist, Ben Lawers NNR	Arctic mouse-ear	*Cerastium nigrescens*
31	25-07-21	Path to mica schist, Ben Lawers NNR	Chestnut rush	*Juncus castaneus*
32	25-07-21	Path to mica schist, Ben Lawers NNR	Alpine mouse-ear	*Cerastium alpinum*
33	25-07-21	Path to mica schist, Ben Lawers NNR	Spiked wood-rush	*Luzula spicata*
34	25-07-21	Path to mica schist, Ben Lawers NNR	Pill sedge	*Carex pilulifera*
35	25-07-21	Path to mica schist, Ben Lawers NNR	Scottish pearlwort	*Sagina x normaniana*
36	25-07-21	Path to mica schist, Ben Lawers NNR	Mossy saxifrage	*Saxifraga hypnoides*
37	25-07-21	Mica schist, Ben Lawers NNR	Alpine poa	*Poa alpina*
38	25-07-21	Mica schist, Ben Lawers NNR	Alpine forget-me-not	*Myosotis alpestris*
39	25-07-21	Mica schist, Ben Lawers NNR	Alpine gentian	*Gentiana nivalis*
40	25-07-21	Mica schist, Ben Lawers NNR	Limestone bedstraw	*Galium sterneri*
41	25-07-21	Mica schist, Ben Lawers NNR	Hairy stonecrop	*Sedum villosum*
42	25-07-21	Mica schist, Ben Lawers NNR	Alpine pearlwort	*Sagina saginoides*
43	25-07-21	Mica schist, Ben Lawers NNR	Alpine saxifrage	*Saxifraga nivalis*

No.	Date	Location	Plant species - common name	Plant species - scientific name
844	25-07-21	Mica schist, Ben Lawers NNR	Mountain sorrel	*Oxyria digyna*
845	25-07-21	Mica schist, Ben Lawers NNR	Alpine fleabane	*Erigeron borealis*
846	25-07-21	Mica schist, Ben Lawers NNR	Rock whitlowgrass	*Draba norvegica*
847	25-07-21	Mica schist, SW of summit, Ben Lawers NNR	Drooping saxifrage	*Saxifraga cernua*
848	25-07-21	Rock outcrop, SE ridge, Ben Lawers NNR	Highland saxifrage	*Saxifraga rivularis*
849	25-07-21	Rock outcrop, bypath on return, Ben Lawers	Net-leaved willow	*Salix reticulata*
850	26-07-21	Roadside N of Cowshill, B6295, N Pennines	Monkshood	*Aconitum napellus*
851	26-07-21	S of ski tow, Yad Moss, County Durham	Marsh saxifrage	*Saxifraga hirculus*
852	26-07-21	Roudsea Wood and Mosses NNR, Cumbria	Giant fescue	*Schedonorus gigantea*
853	26-07-21	Roudsea Wood and Mosses NNR, Cumbria	Large yellow-sedge	*Carex flava*
854	28-07-21	Fivehead Arable Fields, Somerset WT Reserve	Corn parsley	*Petroselinum segetum*
855	28-07-21	Fivehead Arable Fields, Somerset WT Reserve	Broad-leaved spurge	*Euphorbia platyphyllos*
856	28-07-21	Fivehead Arable Fields, Somerset WT Reserve	Slender tare	*Vicia parviflora*
857	28-07-21	Fivehead Arable Fields, Somerset WT Reserve	Corn buttercup	*Ranunculus arvenis*
858	28-07-21	Fivehead Arable Fields, Somerset WT Reserve	Sharp-leaved fluellen	*Kickxia elatine*
859	28-07-21	Fivehead Arable Fields, Somerset WT Reserve	Spreading hedge-parsley	*Torilis arvensis*
860	28-07-21	Fivehead Arable Fields, Somerset WT Reserve	Corn chamomile	*Anthemis arvensis*
861	28-07-21	Causeway Road, nr River Brue, Glastonbury	Soapwort	*Saponaria officinalis*
862	29-07-21	Steps to top, Brean Down, WsM	Common calamint	*Clinopodium ascendens*
863	29-07-21	Steps to top, Brean Down, WsM	White rockrose	*Helianthemum appenin*
864	31-07-21	Borth Bog, Cors Fochno, Dyfi NNR	White beak-sedge	*Rhyncospora alba*
865	31-07-21	Borth Bog (Cors Fochno), Dyfi NNR	Irish lady's-tresses	*Spiranthes romanzoffia*
866	31-07-21	Woodland close to Borth Bog	Sessile oak	*Quercus petraea*
867	31-07-21	N spit, by dunes, Ynyslas, Dyfi NNR	Prickly saltwort	*Salsola kali*
868	31-07-21	Saltmarsh, Ynyslas, Dyfi NNR	Common cord-grass	*Spartina anglica*
869	31-07-21	River Usk, by A40, nr Crickhowell	Himalayan balsam	*Impatiens glandulifera*
870	02-08-21	Behind Bristol Water, nr CV Lake	Horse chestnut	*Aesculus hippocastanum*
871	02-08-21	Disturbed ground, by Salt & Malt, CV Lake	Fat hen	*Chenopodium album*
872	02-08-21	Bristol Water, Wallycourt Road, Chew Stoke	Rowan	*Sorbus aucuparia*
873	02-08-21	Car park, Cranwich Camp, Breckland	Lucerne	*Medicago sativa* ssp. *sat*
874	02-08-21	Meadow, Cranwich Camp, Breckland	Spiny restharrow	*Ononis spinosa*
875	02-08-21	Car park, Cranwich Camp, Breckland	Proliferous pink	*Petrorhagia prolifera*
876	03-08-21	Roadside, West Rudham, N Norfolk	Canadian goldenrod	*Solidago canadensis*
877	03-08-21	Rearing area, Pensthorpe, N Norfolk	Gallant soldier	*Galinsoga parviflora*
878	03-08-21	Wild Ken Hall, Norfolk	Flowering rush	*Butomus umbellatus*
879	03-08-21	Wild Ken Hall, Norfolk	Small nettle	*Urtica urens*
880	04-08-21	Harling Drove, nr Brandon, Breckland	Dark mullein	*Verbascum nigrum*
881	04-08-21	Forest track, Harling Drove, nr Brandon	Red-tipped cudweed	*Filago lutescens*
882	04-08-21	Roadside, A31 S of Winchester	Sainfoin	*Onobrychis viciifolia*
883	04-08-21	Walk down to Cogden Beach, Dorset	Lesser centaury	*Centaurium pulchellum*
884	04-08-21	Walk down to Cogden Beach, Dorset	Clustered dock	*Rumex conglomeratus*
885	04-08-21	Shingle, Chesil Beach, nr Burton Bradstock	Sea couch	*Elymus athericus*
886	04-08-21	Shingle, Chesil Beach, nr Burton Bradstock	Sand couch	*Elymus junceiformis*
887	04-08-21	Shingle, Chesil Beach, nr Burton Bradstock	Autumn hawkbit	*Leontodon autumnalis*
888	04-08-21	Shingle, Chesil Beach, nr Burton Bradstock	Chinese tea-plant	*Lycium chinense*
889	04-08-21	Walk back to car park, Burton Bradstock	Golden melilot	*Melilotus altissima*
890	04-08-21	Walk down to beach at Eype, nr Bridport	Broad-leaved dock	*Rumex obtusifolius*
891	04-08-21	Eroding cliffs at Eype, nr Bridport	Slender centaury	*Centaurium tenuiflorum*
892	04-08-21	Bankside, nr beach, Burton Bradstock	Sea heath	*Frankenia laevis*
893	05-08-21	Path, Hartland Moor NNR, Dorset	Dwarf gorse	*Ulex minor*
894	05-08-21	Wet heath, Hartland Moor NNR, Dorset	Dorset heath	*Erica ciliaris*
895	06-08-21	Wet heath, Stoborough Heath NNR, Dorset	Marsh gentian	*Gentiana pneumonanth*
896	06-08-21	Roadside verge, Soldiers Road, Dorset	Orange day-lily	*Hemerocalis fulva*
897	06-08-21	Middle Beach Cafe, Studland	Montbretia	*Crocosmia x crocosmiifl*
898	07-08-21	Hedgerow nr Woodford Lodge, CV Lake	Walnut	*Juglans regia*

o.	Date	Location	Plant species - common name	Plant species - scientific name
9	07-08-21	Soakaway Pond at CV Lake	Reed canary-grass	*Phalaris arundinacea*
0	07-08-21	Soakaway Pond at CV Lake	Bulrush	*Typha latifolia*
1	08-08-21	Godney Peat Works, Westhay Heath, Levels	Purple loosestrife	*Lythrum salicaria*
2	08-08-21	FP SE of peat works, Westhay Heath	Cockspur grass	*Echinochloa crus-galli*
3	08-08-21	FP to Shapwick Road, Westhay Heath	Water-plantain	*Alisma plantago-aquatica*
4	08-08-21	FP to Shapwick Road, Westhay Heath	Greater burdock	*Arctium lappa*
5	08-08-21	FP to Shapwick Road, Westhay Heath	Fat duckweed	*Lemna gibba*
6	08-08-21	FP E of Shapwick Road, Westhay Heath	Golden dock	*Rumex maritimus*
7	08-08-21	Nr Heathway Drove, Westhay Heath	Water dock	*Rumex hydrolapathum*
8	08-08-21	Roadside ditch, Burtle Road, SW of Westhay	Reed sweet-grass	*Glyceria maxima*
9	08-08-21	Roadside ditch, Burtle Road, SW of Westhay	Frogbit	*Hydrocharis morsus-ranae*
0	09-08-21	Wylye River, Langford Lakes, Wilts WT	Stream water-crowfoot	*Ranunculus penicillatus*
1	09-08-21	E of road, River Wylye, Langford Lakes, Wilts	Gipsywort	*Lycopus europaeus*
2	09-08-21	E of road, River Wylye, Langford Lakes, Wilts	Fool's watercress	*Helosciadium nodoflorum*
3	09-08-21	E of road, River Wylye, Langford Lakes, Wilts	Spiked water milfoil	*Myriophyllum spicatum*
4	09-08-21	E of road, River Wylye, Langford Lakes, Wilts	Water forget-me-not	*Myosotis scorpioides*
5	09-08-21	Path along edge of lake, Langford Lakes	Lesser pond-sedge	*Carex acutiformis*
6	09-08-21	Path along edge of lake, Langford Lakes	Hop	*Humulus lupulus*
7	09-08-21	Water at edge of lake, Langford Lakes	Ivy-leaved duckweed	*Lemna trisulca*
8	09-08-21	Water at lake edge, Langford Lakes	Common duckweed	*Lemna minor*
9	09-08-21	Water at lake edge, Langford Lakes	Yellow water-lily	*Nuphar lutea*
0	09-08-21	Underwater, lake edge, Langford Lakes	Nuttall's waterweed	*Elodea nuttallii*
1	09-08-21	Pond by entrance road, Langford Lakes	Lesser water-parsnip	*Berula erecta*
2	09-08-21	Along road, Langford Lakes	Common comfrey	*Symphytum officinale*
3	09-08-21	Pond by Visitor Centre, Langford Lakes	Rigid hornwort	*Ceratophyllum demersum*
4	09-08-21	Water meadow, Langford Lakes	Common meadow-rue	*Thalictrum flavum*
5	09-08-21	Water meadow, Langford Lakes	Purple willow	*Salix purpurea*
6	09-08-21	Wet woodland edge, Langford Lakes	Hairy-brome	*Bromopsis ramosa*
7	09-08-21	Fisherman's path, Langford Lakes	Green-flowered helleborine	*Epipactus phyllanthes*
8	09-08-21	Fisherman's path, Langford Lakes	Grey sedge	*Carex divulsa*
9	09-08-21	By River Avon, Bradford OA	Greater dodder	*Cuscuta europaea*
0	11-08-21	Yeosden NR, Chilterns	Chiltern gentian	*Gentianella germanica*
1	11-08-21	Yeosden NR, Chilterns	Alsike clover	*Trifolium hybridum*
2	11-08-21	St Mary Church, Radnage, nr Yeosden	Wall lettuce	*Mycelis muralis*
3	11-08-21	St Mary Church, Radnage, nr Yeosden	Traveller's joy	*Clematis vitalba*
4	11-08-21	Barley field, N of Radnage, nr Yeosden	Fool's parsley	*Aethusa cynapium*
5	11-08-21	Road E of Parslow's Hillock, Chilterns	Violet helleborine	*Epipactus purpurata*
6	12-08-21	Heathland pond, Surrey	Alternate water milfoil	*Myriophyllum alterniflorum*
7	12-08-21	Heathland pond, Surrey	French cranesbill	*Geranium endressii*
8	12-08-21	Heathland pond, Surrey	Bulbous rush	*Juncus bulbosus*
9	12-08-21	Heathland pond, Surrey	Water purslane	*Lythrum portula*
0	12-08-21	Heathland pond, Surrey	Sweet chestnut	*Castanea sativa*
1	12-08-21	Reserve, nr Dorking, Surrey	White poplar	*Populus alba*
2	12-08-21	Ponds, nr Dorking, Surrey	Starfruit	*Damasonium alisma*
3	12-08-21	Ponds, nr Dorking, Surrey	Floating sweet-grass	*Glyceria fluitans*
4	12-08-21	Ponds, nr Dorking, Surrey	Cyperus sedge	*Carex pseudocyperus*
5	12-08-21	Flint Hill Roundabout, A24, S of Dorking	Prickly lettuce	*Lactuca serriola*
6	12-08-21	Thorpe Hay Meadow, Surrey WT	Buckthorn	*Rhamnus cathartica*
7	15-08-21	Nr Observatory, Avon Gorge, Bristol	Keeled garlic	*Allium carinatum*
8	15-08-21	Woodland, Yorkshire Sculpture Park	Wild pear	*Pyrus pyraster*
9	15-08-21	Gateway between A1M and M1 J47	Thistle broomrape	*Orobanche reticulata*
0	17-08-21	Loch Inch shoreline, Kincraig	Canadian waterweed	*Elodea canadensis*
1	17-08-21	Old Military Road, Anagach Wood	Male fern	*Dryopteris felix-mas*
2	17-08-21	By Spey River, FP between bridges, GoS	European gorse	*Yulex europaea*

No.	Date	Location	Plant species - common name	Plant species - scientific name
953	17-08-21	By River Spey, FP between bridges, GoS	Common Michaelmas-daisy	*Symphyotrichum x salignum*
954	17-08-21	Along Military Road, GoS	Blue globe-thistle	*Echinops bannaticus*
955	17-08-21	Along Military Road, GoS	Common couch	*Elymus repens*
956	22-08-21	Derelict area, B970, SW of Nethy Bridge	Treacle mustard	*Erysimum cheiranthoi*
957	23-08-21	Along A49 nr J25 M6, Wigan Premier Inn	American willowherb	*Epilobium ciliatum*
958	23-08-21	Waste ground A49, Wigan Premier Inn	Creeping soft-grass	*Holcus mollis*
959	26-08-21	Ditch, Walton Moor, nr Clevedon	Brown galingale	*Cyperus fuscus*
960	29-08-21	Avoncliff, Kennett and Avon Canal	Orange balsam	*Impatiens capensis*
961	29-08-21	Barton Bridge, Barton Farm CP, Bradford OA	Unbranched bur-reed	*Sparganium emersum*
962	29-08-21	Barton Bridge, Barton Farm, Bradford OA	Loddon pondweed	*Potamogeton nodosus*
963	29-08-21	Road down to path, Upper Westood	Balm	*Melissa officinalis*
964	30-08-21	Cross Quarry, nr Axbridge, Mendips	Autumn lady's-tresses	*Spiranthes spiralis*
965	30-08-21	Cross Quarry, nr Axbridge, Mendips	Wall cotoneaster	*Cotoneaster horizonta*
966	30-08-21	A370, Brunel Way, nr Greville Smythe, Bristol	Annual wall-rocket	*Diplotaxis muralis*
967	30-08-21	Island, nr swingbridge, Bristol	Pot marigold	*Calendula officinalis*
968	30-08-21	Island, nr swingbridge, Bristol	Skullcap	*Scutellaria galericulata*
969	30-08-21	Island, nr swingbridge, Bristol	Beggarticks	*Bidens frondosa*
970	30-08-21	Island, nr swingbridge, Bristol	Garden pink-sorrel	*Oxalis latifolia*
971	30-08-21	Island, nr swingbridge, Bristol	Shaggy-soldier	*Galinsoga quadriradia*
972	31-08-21	B3114, Herons Green Bay, CV Lake	Sunflower	*Helianthus annuus*
973	31-08-21	East Harptree Combe, East Harptree	Spreading bellflower	*Campanula patula*
974	01-09-21	Gutter at RV point, Grimley, Worcs	Black nightshade	*Solanum nigrum*
975	01-09-21	Road verge at RV point, Grimley	Procumbent yellow-sorrel	*Oxalis corniculata*
976	01-09-21	Road verge at RV point, Grimley	Crab apple	*Malus sylvestris*
977	01-09-21	FP, next to gravel pits, Grimley Pits, Worcs	Trifid bur-marigold	*Bidens tripartita*
978	01-09-21	FP, next to gravel pits, Grimley Pits, Worcs	Great horsetail	*Equisetum telmateia*
979	01-09-21	Arable fields, nr pits, Worcs	Many-seeded goosefoot	*Chenopodium polysperr*
980	01-09-21	Edge of large pit, Grimley Pits, Worcs	Nodding bur-marigold	*Bidens cernua*
981	01-09-21	Underwater, N edge of pit, Worcs	Fennel pondweed	*Potamogeton pectinatu*
982	01-09-21	Camp Lane, Grimley Village	Dwarf mallow	*Malva neglecta*
983	01-09-21	Field along FP, W of church, Grimley	Common amaranth	*Amaranthus retroflexu*
984	01-09-21	Field along FP, W of church, Grimley	Green field-speedwell	*Veronica agrestis*
985	01-09-21	Field along FP, W of church, Grimley	Pale persicaria	*Persicaria lapathifolia*
986	01-09-21	Field along FP, W of church, Grimley	Red goosefoot	*Chenopodium rubrum*
987	01-09-21	Field along FP, W of church, Grimley	Green nightshade	*Solanum physalifolium*
988	01-09-21	FP, nr Old Vicarage, Grimley	Coriander	*Coriandrum sativum*
989	05-09-21	Path, Priddy Mineries, N of Wells	Slender rush	*Juncus tenuis*
990	05-09-21	Mere, Priddy Mineries, N of Wells	Broad-leaved pondweed	*Potamogeton natans*
991	05-09-21	Ditch, Priddy Mineries, N of Wells	Limestone fern	*Gymnocarpium robertia*
992	05-09-21	Stockhill Woods, N of Wells	Sherard's downy-rose	*Rosa sherardii*
993	05-09-21	Stockhill Woods, N of Wells	Narrow scaly male-fern	*Dryopteris cambrensis*
994	05-09-21	Helena's garden wall, Paulton	Flattened meadow grass	*Poa compressa*
995	08-09-21	Nr Burrow Mump, Southlake Moor NNR	Arrowhead	*Sagittaria sagittifolia*
996	08-09-21	Burrow Hill Drove, Southlake Moor NNR	Marshmallow	*Althaea officinalis*
997	08-09-21	A372 verge, E of Bridgwater, by M5	Slender hare's-ear	*Bupleurum tenuissimu*
998	08-09-21	Stolford coast, Bridgwater Bay	Sea wormwood	*Atemisia maritima*
999	15-09-21	Clifftop, Cheddar Gorge	Somerset hair-grass	*Koeleria vallesiana*
1000	18-09-21	Top FP, Uphill NR, Uphill	Goldilocks aster	*Galatella linosyris*
1001	19-08-21	Grassland along River Parrett, Combwich	Hard-grass	*Parapholis strigosa*
1002	19-08-21	Grassland along River Parrett, Combwich	Common saltmarsh-grass	*Puccinellia maritima*
1003	19-08-21	Grassland along River Parrett, Combwich	Annual sea-blite	*Suaeda maritima*
1004	19-08-21	Cricket pitch, Combwich	Equal-leaved knotgrass	*Polygonum arenastrum*
1005	19-08-21	Cricket pitch, Combwich	Bulbous foxtail	*Alopecuris bulbosus*
1006	19-08-21	Grassland along River Parrett, Combwich	Reflexed saltmarsh-grass	*Puccinellia distans*

o.	Date	Location	Plant species – common name	Plant species – scientific name
07	19-08-21	Grazing marsh, FP, SE of Steart Marshes (WWT)	Black bent	*Agrostis gigantea*
08	19-08-21	Rhyne on grazing marsh, Steart Marshes (WWT)	Grey club-rush	*Schoenoplectus tabernaemontani*
09	19-08-21	Ditch across grazing marsh, Steart	Marsh foxtail	*Alopecuris geniculatus*
10	19-08-21	Ditch across grazing marsh, Steart	Yellow bristle-grass	*Setaria pumila*
11	19-08-21	Grazing marsh, close to the River Parrett	Canarygrass	*Phalaris canariensis*
12	19-08-21	Hedgerow, nr River Parrett	English elm	*Ulmus procera*
13	19-08-21	Ditch across grazing marsh, Steart	Greater duckweed	*Spirodela polyrhiza*
14	19-08-21	Ditch on grazing marsh, Steart	Wild celery	*Apium graveolens*
15	25-09-21	Brimble Pit Pool, nr Westbury-sub-Mendip	Ivy-leaved crowfoot	*Ranunculus hederaceus*
16	25-09-21	Brimble Pit Pool, nr Westbury-sub-Mendip	Horned pondweed	*Zannichellia palustris*
17	28-09-21	Veg beds in Community Farm, Chew Stoke	Wall speedwell	*Veronica arvensis*
18	28-09-21	Veg beds in Community Farm, Chew Stoke	Fig-leaved goosefoot	*Chenopodium ficifolium*
19	28-09-21	Veg beds in Community Farm, Chew Stoke	Hoary willowherb	*Epilobium parviflorum*
20	28-09-21	Veg beds in Community Farm, Chew Stoke	Short-fruited willowherb	*Epilobium obscurum*
21	28-09-21	Polytunnel, Community Farm, Chew Stoke	Northern yellow-cress	*Rorippa islandica*
22	28-09-21	Polytunnel, Community Farm, Chew Stoke	Square-stemmed willowherb	*Epilobium tetragonum*
23	28-09-21	Polytunnel, Community Farm, Chew Stoke	Bilbao fleabane	*Erigeron floribunda*
24	28-09-21	Hedgerow, Community Farm, Chew Stoke	Pedunculate oak	*Quercus robor*
25	29-09-21	Coombe Hill, Chilterns	Fringed gentian	*Gentianopsis ciliata*
26	30-09-21	Arable field, nr Glandford, Norfolk	Flixweed	*Descurainia sophia*
27	30-09-21	Arable field, nr Glandford, Norfolk	Grey field-speedwell	*Veronica polilta*
28	30-09-21	Arable field, nr Glandford, Norfolk	Common fiddleneck	*Amsinckia micrantha*
29	30-09-21	Morston Quay saltmarsh NT, nr Blakeney	Shrubby sea-blite	*Suaeda vera*
30	30-09-21	Morston Quay saltmarsh NT, nr Blakeney	Purple glasswort	*Salicornia ramosissima*
31	30-09-21	Morston Quay saltmarsh NT, nr Blakeney	Sea rush	*Juncus maritimus*
32	30-09-21	Morston Quay saltmarsh NT, nr Blakeney	One-flowered glasswort	*Salicornia disarticulata*
33	30-09-21	Morston Quay saltmarsh NT, nr Blakeney	Long-bracted sedge	*Carex extensa*
34	30-09-21	Morston Quay saltmarsh NT, nr Blakeney	Long-spiked glasswort	*Salicornia dolichostachya*
35	30-09-21	Morston Quay saltmarsh NT, nr Blakeney	Yellow glasswort	*Salicornia fragilis*
36	30-09-21	Morston Quay saltmarsh NT, nr Blakeney	Perennial glasswort	*Sarcocornia perennis*
37	30-09-21	Field by disused railway, Wighton	Maple-leaved goosefoot	*Chenopodiastrum hybridum*
38	30-09-21	Field, E of Flitcham	Night-flowering catchfly	*Silene noctiflora*
39	30-09-21	Field, E of Flitcham	Cut-leaved dead-nettle	*Lamium hybridum*
40	30-09-21	Nr pumping station, Abbey Road, Flitcham	Hoary mullein	*Verbascum pulverulentum*
41	30-09-21	CP for church, Edgefield	Small toadflax	*Chaenorhinum minus*
42	30-09-21	Crossroads, SW of Edgefield	Orpine	*Hylotelephium telephium*
43	09-10-21	Velvet Bottom Somerset WT	Meadow saffron	*Colchicum autumnale*
44	14-10-21	Lucky Lane, Southville, Bristol	Argentinian vervain	*Verbena bonariensis*
45	30-10-21	By Birnbeck Road, WsM	Water bent	*Polypogon viridis*
46	30-10-21	By Grand Pier viewpoint, WsM	Moroccan eryngo	*Eryngium variifolium*
47	30-10-21	By car park, Raglan Place, WsM	Mind-your-own-business	*Soleirolia soleirolii*
48	30-10-21	By Lisbon Hotel, Greenfield Place, WsM	Tree-of-heaven	*Ailanthus altissima*
49	30-10-21	By Lisbon Hotel, Greenfield Place, WsM	Cider gum	*Eucalyptus gunnii*
50	30-10-21	Seafront promenade, WsM	Guernsey fleabane	*Erigeron sumatrensis*
51	30-10-21	Marine Promenade, WsM	Cabbage palm	*Cordyline australis*
52	30-10-21	Marine Promenade, WsM	Silver ragwort	*Jacobaea maritima*
53	30-10-21	Marine Promenade, WsM	Intermediate polypody	*Polypodium interjectum*
54	30-10-21	Roundabout, by Marine Promenade, WsM	Toothed medick	*Medicago polymorpha*
55	30-10-21	Roundabout, by Marine Promenade, WsM	Bermuda-grass	*Cynodon dactylon*
56	30-10-21	Roundabout, by Marine Promenade, WsM	Rescue brome	*Ceratochloa cathartica*
57	30-10-21	Winter Gardens Pavillion, WsM	Fiddle dock	*Rumex pulcher*
58	20-11-21	Lossiemouth Forest, Moray coast	Himalayan cotoneaster	*Cotoneaster simonsii*
59	27-12-21	Road by Little Trevothan Campsite, Lizard	Altar-lily	*Zantedeschia aethiopica*

Index